PREPARING TEACHERS FOR
Three-Dimensional Instruction

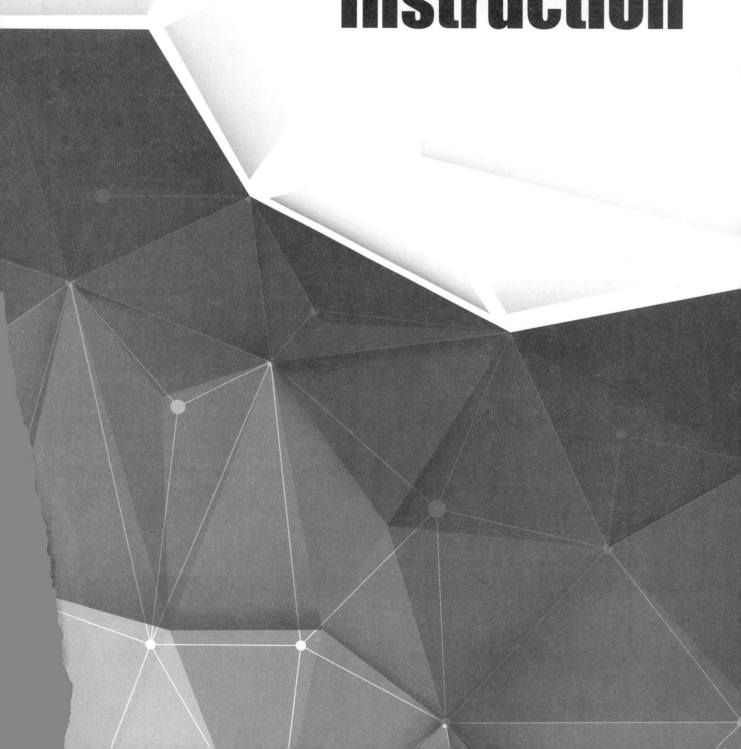

PREPARING TEACHERS FOR
Three-Dimensional Instruction

Jack Rhoton, Editor

NSTApress

National Science Teachers Association

Arlington, Virginia

National Science Teachers Association

Claire Reinburg, Director
Rachel Ledbetter, Managing Editor
Deborah Siegel, Associate Editor
Andrea Silen, Associate Editor
Donna Yudkin, Book Acquisitions Manager

Art and Design
Will Thomas Jr., Director
Cover and interior design, Jae Martin

Printing and Production
Catherine Lorrain, Director

National Science Teachers Association
David L. Evans, Executive Director

1840 Wilson Blvd., Arlington, VA 22201
www.nsta.org/store
For customer service inquiries, please call 800-277-5300.

All photos are courtesy of the chapter authors unless otherwise noted.

Cataloging-in-Publication Data for this book and the e-book are available from the Library of Congress.
ISBN: 978-1-68140-393-9
e-ISBN: 978-1-68140-394-6

The Next Generation Science Standards ("NGSS") were developed by twenty six states, in collaboration with the National Research Council, the National Science Teachers Association and the American Association for the Advancement of Science in a process managed by Achieve, Inc. more information go to *www.nextgenscience.org.*

Contents

SECTION 1

Shifts in Teacher Knowledge and Practice: Models of Teaching to Meet the Intent of the *NGSS*

Contents

Contents

Contents

Foreword

Bruce Alberts

The recent U.S. election cycle has made scientists strikingly aware of the potential dangers that society faces when a major segment of the population no longer accepts findings that have been arrived at through scientific consensus. Currently, a widespread distrust of experts and a preference for "alternative facts" threatens democracies. This realization raises the stakes for our education systems: We urgently need to provide future generations with a much-improved science education that empowers adults to function as effective problem solvers and make wise decisions for themselves, their families, and their nation.

I especially admire the clear recognition in this volume that we are all in this together. Both the college faculty who teach undergraduate science classes and the K–12 science teachers in our nation's schools must make substantial changes not only to the goals that they have for their students but also to the pedagogies that they use to teach them. And at all levels, science education cannot primarily be considered an effort to produce more scientists, as it was in the post-*Sputnik* years, when the United States felt technologically threatened by the Soviet Union. Instead, our main goal must be to produce citizens who can "think like a scientist," basing the many important decisions that they must make in their lives on evidence and logic rather than on emotions and "magical thinking."

The challenge in education today is how to produce adults who are effective, rational thinkers able to contribute to (and cope with) our complex and constantly changing societies. To have any chance of success, science courses, including those at the postsecondary level, must discard the all-too-frequent aim of "covering" an entire field like physics, chemistry, Earth sciences, or biology. Only in this way can teachers provide their students with the time they'll need to delve deeply into a few scientific problems and actually employ scientific reasoning based on evidence. Students will also need to learn how scientific knowledge is built up over time through the combined efforts of thousands of independent scientists, enabling them to appreciate the consensus views of the scientific community on issues such as climate change and vaccine safety. Thus, conveying the essence of "science as a way of knowing" needs to become an explicit goal for all science courses.

Consider my own field of biology. Although scientific study of life on Earth should be especially fascinating for all students, this area of scientific research has accumulated such a vast amount of knowledge that introductory courses are often rendered almost useless by attempts to "cover" all that is known. Because of such demands, I find the textbooks used to introduce biology at every level incredibly dull. Even in a very large book, there is rarely enough space to reflect what's really exciting about any one subject. Instead of merely being given the answer to an important biological problem, students should first be forced to struggle with the same challenges that puzzled the best scientists before they were able to solve it.

If we are to produce adults who can use scientific thinking effectively in their daily lives, science teachers at all levels must greatly reduce the number of facts that we cover in class. The *Next Generation Science Standards* (NGSS Lead States 2013) call on K–12 teachers to teach in a manner very different from the way most of them were taught in college. As a result, huge amounts of teacher professional development will be needed to create the type of science education that we seek. Only by teaching very differently at the college level can we ever hope to introduce the same shifts in K–12 classrooms. We are all truly in this together, as this book directly demonstrates.

When I was unexpectedly selected to become the full-time president of the National Academy of Sciences in 1993, I resisted at first because it would require that

Foreword

I close my research laboratory at UCSF. The Academy's Nomination Committee was finally successful in recruiting me to Washington, D.C., only because they promised that I could become an "education president." Over the course of my 12-year presidency (1993–2005), the academy would in fact publish well over a hundred reports on education, including the first-ever *National Science Education Standards* (NRC 1996). We also partnered vigorously with the Smithsonian Institution to spread inquiry-based elementary and middle school science education through workshops and curricula produced by the National Science Resources Center (now the Smithsonian Science Education Center). Some members of the academy disagreed with this focus, claiming that science education at the K–12 level had nothing to do with postsecondary teacher education but was rather the job of textbook publishers, school boards, and teachers unions. As the contents of this book demonstrate, these critics were clearly quite wrong.

Bruce Alberts *is the Chancellor's Leadership Chair for Science and Education at the University of California, San Francisco. He is the former Editor-in-Chief of Science and President Emeritus of the National Academy of Sciences. He can be reached by e-mail at* balberts@ucsf.edu.

References

National Research Council (NRC). 1996. *National Science Education Standards.* Washington, DC: National Academies Press.

NGSS Lead States. 2013. *Next Generation Science Standards: For states, by states.* Washington, DC: National Academies Press. *www.nextgenscience.org/next-generation-science-standards.*

Preface

The arrival of the *Next Generation Science Standards* (*NGSS*; NGSS Lead States 2013) emanating from the vision established in *A Framework for K–12 Science Education* (*Framework*; NRC 2012) has changed the conversation about best practices for science teaching and learning in our nation's classrooms. This renewed dialogue has not only permeated K–12 science education but also heightened the conversation around changes that need to take place in higher education, both in the experiences that preservice science teachers receive in the science disciplines and in their teacher education programs. These changes extend to designing professional development strategies that help established science teachers understand and embrace the *NGSS*. State policies and regulations will also influence both the pace of implementation of the standards and the manner and fidelity with which they are implemented. The *NGSS* will affect all areas of teaching and learning, including "curriculum, teacher development, and assessment and accountability measures" (Bybee 2014, p. 214). The ultimate goal of these changes is to improve student learning.

The recommendations of the *Framework* and the *NGSS* are shaped by years of research on teaching and learning (NRC 2007). Central to the vision of both is the integration of disciplinary core ideas, science and engineering practices, and crosscutting concepts, which is referred to as three-dimensional learning. In order to achieve this goal, many science teachers will need to change their instructional approaches (Banilower et al. 2013). Chief among the changes is the need for educators to move away from teaching isolated facts and toward practices that engage students in building models through investigations, asking questions, finding solutions to problems, and making sense of phenomena (NRC 2012).

Several states have already adopted the *NGSS*, and additional states are expected to follow. Additionally, some of the states that have chosen not to adopt the new standards in their entirety have adapted much of the language and vision of the *NGSS* into their own state science standards. The vision laid out in the *Framework* and the new science standards has already begun to influence the way science is taught in our nation's schools and will likely drive science education reform for decades to come, affecting the vast majority of science students in the United States.

The purpose of this book is to showcase some instructional approaches that instructors at all levels of education are using to unlock the vision of these new standards, as well as the ways in which both preservice and established teachers are being trained. At the heart of this book is an attempt to showcase shifts, some incremental in nature, that are being made by K–12 science teachers, higher education science faculty, teacher education faculty, other science educators, and policy makers in order to implement the new standards. Fully implementing the vision of the *NGSS* will be a daunting, complex, and time-consuming task. There is no magic wand for achieving the vision. Instead, educators will need to apply a variety of approaches and efforts over an extended period of time. If these new standards are to be implemented with fidelity, we must be willing to shift educational experiences away from the often formulaic methods that many students are now experiencing and support them as they take on the role of practitioners of science.

The science education community is demonstrating that it is ready to meet the challenge of implementing the *NGSS*. In order to do so, science teachers will need to know more than just the proficiency expectations. Merely reading through the new standards and correlating their content to an established curriculum is not sufficient. Preparation for implementing the *Framework*'s vision of teaching and learning will be most effective if it begins during undergraduate coursework and is continuously

Preface

supported through an effective professional development agenda designed to bring about real change in the classroom. Even teachers with years of experience will benefit from extensive support and professional learning experiences, as well as from receiving constructive feedback on what it means to develop motivating lessons and engage students in experiments and investigations that allow them to make sense of core concepts and identify relationships among ideas.

In this volume, we consider a broad range of highly significant topics that affect science teachers and science teacher preparation. We discuss strategies and approaches used by science instructors at various levels, including K–12 science teachers, teacher education faculty, and undergraduate science faculty, and that align with the vision of the *Framework* and the *NGSS*. We also describe professional development programs and approaches for updating practice to meet the goals established in the *NGSS*. Additionally, we discuss the need for state policies and regulations to level the playing field for teachers so that the *NGSS* can be implemented on a wider scale, and we offer examples of successful partnerships among K–12 education, higher education, businesses, and informal science education that can be addressed through STEM education.

The eighteen chapters in this book are organized into five major sections, each with a general theme. The intent is to show how practitioners at various levels are beginning to capture the vision of the *Framework* and the *NGSS*. We believe that the contents of this volume will serve as a motivating resource for the science education community that helps them to harness skills, expertise, and passion as they look to revitalize science instruction. The major themes of this book are highlighted below.

Shifts in Teacher Knowledge and Practice: Models of Teaching to Meet the Intent of the *NGSS*

Teachers represent the critical link between the curriculum and students. Their ability to effectively engage students in three-dimensional learning is key to helping students "think like a scientist." This section describes the changes that some of the nation's outstanding science teachers are making in their classrooms to address the vision called for in the *Framework* and the *NGSS*, such as shifting from teaching science as inquiry to teaching science as practice. Science and engineering practices, crosscutting concepts, and disciplinary core ideas are addressed in this section, as are issues related to *NGSS*-aligned curriculum planning and methods of assessment.

Professional Development Strategies That Support the Implementation of the *Framework* and the *NGSS*

Teachers and administrators must recognize that student learning, classroom teaching, curriculum materials, and student assessment all play a role in ongoing professional development that promotes effective implementation of the *NGSS*. This section describes examples of professional development strategies for helping K–12 science teachers address specific subject matter while learning more about instructional activities that have been proven effective at promoting critical thinking and depth of understanding. These strategies show teachers how to develop content-specific learning opportunities that allow students to engage in argumentation and develop science models in the context of their science lessons rather than in isolation. The type of professional development discussed in this section offers many opportunities for active learning and problem solving in the classroom.

Teacher Preparation Courses

This section discusses the ways in which education faculty are supporting future teachers' understanding of disciplinary core ideas, science and engineering practices, and crosscutting concepts. As teacher education programs adjust to the shifts called for in the *Framework* and the *NGSS*, preservice teachers need opportunities to build capacities and demonstrate their knowledge as they construct explanations, analyze and interpret data, develop models, and engage in argumentation. Future teachers need to be involved in designing lessons, assessing students, implementing strategies, evaluating outcomes, and reflecting with expert guidance on both the content they are learning and the most effective learning opportunities for students. This section also outlines the state and district policies necessary for smooth implementation of the *NGSS*.

Undergraduate Science Courses for Prospective Teachers

In their undergraduate science courses, preservice teachers see science teaching in action and gain the confidence and skills needed to implement educational shifts within their own classrooms. For this reason, teaching candidates need to experience models of context and content in their undergraduate science courses and in their teacher education preparation programs that fit within the elements of the *Framework* and the *NGSS*. In this section, higher education instructors describe some of the changes they have made to ensure that this is the case, as well as some of the challenges they have confronted. We cannot expect future science teachers to embrace the paradigm shift called for in the *Framework* and the *NGSS* if they have never seen these models applied. Building stronger partnerships between disciplinary departments and schools of education in colleges and universities will help all students attain higher standards.

Harnessing the Business Community and Other Entities to Support the Vision of the *NGSS*

The last section in this volume describes partnerships between industry and education to achieve common objectives aligned to the *Framework* and the *NGSS*. Successful partnerships ensure that K–12 students receive the skills and knowledge they'll need to succeed in both higher education and the workforce, emphasize the relationship between classroom learning and science careers, and support teachers as they strive to build stronger content knowledge and pedagogical expertise. Specifically, this section explores the work of East Tennessee State University's Center of Excellence in Mathematics and Science Education as well as the ETSU Northeast Tennessee STEM Hub. The section also provides guidelines that leaders can use to form new partnerships, establish common goals, and encourage continuing contributions to meeting those goals.

References

Banilower, E., P. S. Smith, I. R. Weiss, K. A. Malzahn, K. M. Campbell, and A. M. Weiss. 2013. *Report of the 2012 national survey of science and mathematics education.* Chapel Hill, NC: Horizon Research.

Bybee, R. W. 2014. NGSS and the next generation of science teachers. *Journal of Science Teacher Education* 25 (27): 211–221.

National Research Council (NRC). 2007. *Taking science to school: Learning and teaching science in grades K–8.* Washington, DC: National Academies Press.

National Research Council (NRC). 2012. *A framework for K–12 science education: Practices, crosscutting concepts, and core ideas.* Washington, DC: National Academies Press.

NGSS Lead States. 2013. *Next Generation Science Standards: For states, by states.* Washington, DC: National Academies Press. *www.nextgenscience.org/ next-generation-science-standards.*

Acknowledgments

My interest in STEM (science, technology, engineering, and mathematics) education can be traced to early experiences growing up on a farm in rural Scott County, Virginia, during the 1950s and 60s. These experiences ranged from animal husbandry to machinery maintenance. When equipment broke or malfunctioned, it had to be repaired in order for the work to continue. Because replacement parts were not always readily available, creative problem solving, ingenuity, and mechanical skills were essential to bringing the broken farm implements back to functionality. One can argue that such experiences, obtained at an early age, are akin to the three-dimensional learning called for by *A Framework for K–12 Science Education* (*Framework*; NRC 2012) and the *Next Generation Science Standards* (*NGSS*; NGSS Lead States 2013). At a rudimentary level and without knowing it, I was integrating all three dimensions—disciplinary core ideas, science and engineering practices, and crosscutting concepts—on the farm to better explain phenomena and solve problems.

When the Soviet Union launched the *Sputnik* space satellite in October 1957, my interest in science and technology was further galvanized. I have distinct memories of crawling out of my bedroom window and onto the roof late at night to lie on my back and watch *Sputnik* glide through the sky. I was 14 years old and a freshman at Rye Cove High School in Clinchport, Virginia. It was as though I had a front-row seat to the beginning of the space age. The excitement surrounding the *Sputnik* launch changed by life, and in subsequent years I experienced the deep influence that this event had on STEM education in our nation's schools.

I mention these early experiences because I am so grateful for the K–12 teachers who nurtured my interest in STEM. These educators actively engaged their students in the process of learning science. Similarly, I am indebted to the higher education science and science education professors who instilled in me their passion for teaching and learning. In particular, I want to thank Franklin Robinson of Hiwassee College; the late William "Bill" Pafford and Hubert Armantrout of East Tennessee State University; the late Ertle Thompson of the University of Virginia; Daniel Sonenshine of Old Dominion University; and the late A. Paul Wishart of the University of Tennessee. These caring and compassionate scientists and teachers modeled best teaching practices and provided me with life-changing experiences. A host of other individuals on the national stage, too numerous to mention here, has also influenced me and my work in STEM education.

I also wish to express my sincere appreciation to the individuals who made this publication possible. The anonymous reviewers who assessed the book proposal and offered valuable feedback early on in the process have my gratitude. Thanks as well to Gerry Madrazo for his encouragement and assistance.

I would especially like to thank the authors who contributed to this book. No volume is any better than the chapters within it, and I appreciate the time and efforts of those whose work you'll find here. I believe the team of authors assembled to write this book provides a blend of knowledge and skills that contributes uniquely to its purpose. In particular, I would like to thank Amy Selman, my former graduate assistant, for applying her expert guidance to each manuscript. She went above and beyond the call of duty in providing helpful feedback. I also wish to thank my former colleagues at the East Tennessee State University Center of Excellence in Mathematics and Science Education for their continued support and encouragement.

Finally, I am deeply appreciative of the support and assistance provided by the outstanding staff at NSTA Press, including Rachel Ledbetter, Claire Reinburg, and Donna Yudkin.

Acknowledgments

References

National Research Council (NRC). 2012. *A framework for K–12 science education: Practices, crosscutting concepts, and core ideas.* Washington, DC: National Academies Press.

NGSS Lead States. 2013. *Next Generation Science Standards: For states, by states.* Washington, DC: National Academies Press. *www.nextgenscience.org/ next-generation-science-standards.*

1

Shifts in Teacher Knowledge and Practice: Models of Teaching to Meet the Intent of the *NGSS*

CHAPTER 1

Implementing Pedagogical Approaches to Support Students in Science Practices

Kenneth L. Huff

In this chapter, we establish a contemporary view of science practices. We begin by providing a brief historical perspective of the role inquiry has played in the emergence of current practices. Then, we discuss how phenomenon-based investigations help students apply science and engineering practices to meeting expectations outlined in the *Next Generation Science Standards* (*NGSS*; NGSS Lead States 2013). Throughout the chapter, we offer classroom examples of ways to foreground effective science practices during instruction.

View of Science Practices: Effects on Contemporary Classrooms

The importance of inquiry has been a focal point in science education for decades. Schwab (1962) viewed inquiry as a process through which students confront phenomena to learn. In his view, students in inquiry classrooms use evidence to assess the validity of claims made by others. In today's classrooms, focus has shifted to the importance of science practices. *A Framework for K–12 Science Education* (*Framework*; NRC 2012a) emphasizes engagement in science inquiry by requiring students to simultaneously interpret knowledge and demonstrate skills. Contrary to the perceptions of some educators, the *Framework* does not recommend doing away with inquiry. On the contrary, it promotes enriching science instruction by enhancing the role of inquiry in instruction and clarifies the range of intellectual, social, and mental processes such inquiry requires.

The outcomes of students' science practice in the classroom form the bedrock of further discussion, with students developing explanations based on evidence. Although there has been a shift toward science practices

over the years as research on effective instructional strategies has emerged (NRC 2000; NRC 2005; NRC 2007), too many of the activities in schools today do not reflect these advances.

In today's science classrooms, students are expected to engage in science practices directly themselves, not merely learn about them secondhand. *Ready, Set, Science!* (NRC 2008) describes methods for integrating practice with learning about concepts, emphasizing that the two processes cannot be separated. Activities that merge conceptual instruction with student practice through active learning have been proven to increase student proficiency.

In the last decade, I have moved away from the classic approach of limiting inquiry to investigating hypotheses. Teaching the scientific method in an algorithmic manner removes creativity, and I have found that deep, personal reflection is necessary for science instruction to be effective. Instead, I have embraced the idea of students making sense of findings, using data to guide reasoning, and arguing competing explanations to reach consensus. Recommendation 5 in the *Framework* states that performance expectations should require students to demonstrate knowledge through the use of science and engineering practices. For this to happen, teachers must do more than merely present and assess science content. Actual science is much more than a rote answer arrived at long ago by a famous scientist; students need to actively engage with practices as they develop and apply scientific ideas.

When educators embrace science practices in the classroom, they must also educate parents about related conceptual shifts. Here's an example. At a recent parent-teacher conference, the mother of a child with an individualized education program (IEP) asked me if classroom learning would include hands-on experiences, noting this approach helped her child to retain information. We discussed her interpretation of the term *hands-on* and my use of practices for classroom learning. I explained that my course emphasizes asking questions, developing and using models, constructing explanations, and engaging in scientific argumentation. I explained that by engaging students in these practices, I would be able to describe her child's performance in relation to specific activities rather than merely offering generalities. We discussed the value of teaching students to think and reason critically, and the implications of the *Framework*'s vision for science education. In particular, I explained to the parent that I am now less concerned about my students' ability to complete worksheets or memorize facts than I am about encouraging them to develop evidence-based explanations that they can ably critique and argue. At the end of the conference, we agreed that active science practice promotes a deeper level of learning and understanding.

Unfortunately, content learning and oversimplified classroom practices are often the sole foci of instruction. Alberts (2010) argues that science education must center on evaluation of evidence and active participation in science practices. The beauty of the eight practices in the *Framework* and the *NGSS* is they describe distinct and specific abilities related to science inquiry and delineate in detail the processes scientists use to better understand the world around them.

Introducing Phenomena for the Purposeful Use of Science Practices

The *NGSS* become meaningful as instruction shifts from learning about concepts and processes to developing scientific explanations. Student-generated questions cultivate exercises that help learners make sense of observable phenomena and require them to apply science and engineering practices. The New York State Teaching Standards conclude that in today's learning environment, effective pedagogical approaches require teachers to use a variety of technological tools, techniques, and skills to promote student learning. Teachers must also be responsive to students' individual learning needs, strengths, interests, and experiences, and they must provide supports that enable all learners to engage in science and engineering practices. Realizing these goals requires teachers to introduce phenomenon-based investigations in the classroom.

Phenomena pervade the *NGSS*, and performance expectations at the elementary, middle, and high school levels lead students to develop ideas and skills that explain phenomena central to science disciplines. Students experience phenomena when they come to know it through their senses. Though a phenomenon can be mistakenly interpreted as an anomaly (e.g., observing a rainbow), students quickly learn that phenomena don't appear out of nowhere. Once students understand that phenomena have causes, they can seek evidence to support possible explanations for those causes. The Research and Practice Collaboratory recommends the following criteria when introducing phenomena:

1. Build on students' everyday experiences. Phenomena derived from the community (past or present) are often compelling to students because they focus on group identity. Students can address such issues as equity and diversity in the classroom by exploring these phenomena.

2. Because good phenomena are events or processes that students can observe, technological tools such as videos, probes, telescopes, and microscopes can help students to explore the crosscutting concept of patterns.

3. Apply multiple science and engineering practices. Engaging students in a variety of practices is fundamental to nurturing habits of mind essential for science literacy.

Teachers can draw students into investigations by encouraging them to generate ideas and questions by connecting the topic being investigated to real-life experiences (Yager 2006). Emphasizing connections makes learning science more relevant to students, which in turn leads them to produce higher quality work. By relating content to learners' daily lives, educators also encourage students to make interdisciplinary connections and see familiar objects and occurrences in new ways. Providing opportunities for students to investigate, understand, and use objects and events that are of personal or local interest motivates students to ask questions, carry out investigations, and use evidence to support the explanations they develop.

Consider the following example. For my students, algal blooms in New York lakes represent a meaningful phenomenon. This toxic problem, which affects almost every facet of life along the Great Lakes and has become increasingly worse over the past 10 years, has had a devastating effect on tourism and recreation in the area. To begin our investigation of the subject, I present students with a brief description of algal blooms and photographs of its effects. Then, I ask them how they would explain the situation. During our initial dialogue, students wonder about possible factors and seem perplexed. Most students believe the blooms must be caused by some kind of chemical in the water, but they have no idea which one or how it was introduced to the lake.

Next, students work in groups to define the system under study. They formulate additional questions and gather information from various sources about potential causes of the algal bloom. As students formulate questions, I provide instruction helping them identify explicit causal relationships. At this point in the sequence, it is important to devote time to teaching students habits of mind that enable them to consistently formulate questions about causal relationships, because this habit is a key element in making sense of phenomena. In the classroom, this involves nurturing a sense of wonderment and seeking explanations for causes of phenomena students encounter.

Constructing an explanation for the cause of the algal bloom is next in the learning sequence. Students use available resources, including tablets provided by the district, to seek out explanations. They are presented with a spreadsheet containing Lake Erie algal bloom data acquired through the National Oceanic and Atmospheric Administration (NOAA) with the help of a local university. During this portion of the learning sequence, students create a Venn diagram showing the relationship between correlation and causation. Correlations between variables do not imply that a change in one variable is the cause of a change in values of another variable. Causation, however, does indicate one event is the result of the occurrence of the other event. Although students initially are overwhelmed by the complexity and comprehensiveness of the spreadsheet data, after a bit of prompting, they begin to interpret patterns. Students are thus able to identify factors showing causation and correlation, including the positive correlation between total phosphorous and total suspended solids.

Explicit development of crosscutting concepts occurs at this juncture because they contribute to the sense making of novel phenomena. Crosscutting concepts help provide students with an organizational framework for connecting knowledge from various disciplines and serve as powerful tools for thinking and analyzing. Understanding science means comprehending what is common among disciplines (i.e., how to connect ideas being learned in one area with ideas learned previously in another area).

Turning crosscutting concepts into questions and using them to link to practices is a highly successful strategy. For example, as my students examine the algal bloom data, asking themselves what patterns and relationships among features they notice and what further observations might help to clarify patterns and their causes can be an effective learning strategy.

Crosscutting concepts should become common and familiar elements across all science disciplines and grade levels. According to the *Framework*, explicit reference to the concepts and using them in multiple disciplinary contexts can help students develop a cumulative, coherent, and usable understanding of science and engineering (NRC 2012a).

In mathematics, one contemporary approach to inquiry is that of facilitating student development of models used to make sense of quantities and their relationship to a particular phenomenon. By using spreadsheet data to develop these models, students achieve a deeper level of understanding than if they were to simply apply a ready-made model. This approach is effective in both science and mathematics, and it integrates the two disciplines in a meaningful way. After developing algal bloom models, my students use lab books to write an explanation of correlational patterns in the system and support their explanations with evidence from the data.

Whether or not your state adopts the *NGSS*, using phenomena to engage students in science and engineering practices is prudent and effective. These practices include formulating and utilizing evidence to develop, refine, and apply explanations to construct accounts of scientific phenomena (NRC 2012a). These types of experiences help students see their classroom practices as science in action.

Pulling Science Practices to the Foreground

Bybee (2013) refers to content that is emphasized in a lesson as foreground content and content that provides the lesson's context as background content. As a class proceeds through an instructional sequence, such as the 5E Instructional Model (engage, explore, explain, elaborate, evaluate), different practices can be either in the foreground or background.

I have found that pulling certain practices to the foreground during instruction helps to make them explicit. During lessons, I frequently stop to remind students of the science and engineering practice we are emphasizing and the reason we have brought it to the foreground. Once students become more familiar with science practices, I transfer the responsibility of identifying the purpose of emphasizing different ones to students. Although instruction integrates all three dimensions of science learning—science and engineering practices, crosscutting concepts, and disciplinary core ideas—it is essential for both teacher and students to be aware of which particular dimension is being emphasized at a given point in the instructional sequence.

Incorporating science and engineering practices into classroom exercises is only one of the instructional shifts required to transition to an *NGSS* classroom. Educators must also align practices with performance expectations that best facilitate comprehension of the concept being investigated. This goal can be achieved through use of backward design (Wiggins and McTighe 2005). Investigating the Lake Erie algal bloom phenomenon, I began by identifying essential learning outcomes of the unit and then engaged students in a 5E storyline sequence leading to those outcomes.

Using backward design, I divided the planning processes into stages. In stage one, I selected two core ideas from the *Framework*, specifically from LS2.C (Ecosystem Dynamics, Functioning, and Resilience) and ESS3.C (Human Impacts on Earth Systems).

I used these core ideas to construct a Lake Erie algal bloom storyline that would help students reach a desired level of proficiency. Krajcik et al. (2014) assert that storylines show the ongoing development of disciplinary core ideas, crosscutting concepts, and science and engineering practices. Moreover, they help students build sophisticated ideas from prior experiences and allow them to develop evidence leading to conceptual understanding. When students use storylines to engage in science and engineering practices, I have found that they develop explanations of phenomena that either meet or exceed performance expectations.

In stage two, I determined ways for students to demonstrate attainment of the desired level of understanding. It is important to do more than merely expose students to material. Teachers must be able to determine whether their students are achieving the intended learning. During this phase, I did not define algal blooms or nonpoint source pollution. Instead, I assessed students' knowledge by listening to them discuss algal blooms, the factors that correlate with the presence of phosphates, and the thought processes they used to determine the condition of the lake ecosystem. By listening to students, I verified that they were able to analyze Lake Erie algal bloom data and construct explanations for causal and correlational relationships. Through dialogue, students demonstrated whether they met the goal of using evidence to support explanations. Many students also shared arguments supporting the relevance of specific evidence to their explanations.

In stage three, I identified the content of the core (background) ideas and determined crosscutting concepts and science and engineering practices to pull to the foreground. I used this information to develop learning experiences for each of the phases of the 5E Instructional Model: engage, explore, explain, elaborate, and evaluate. We used the NOAA spreadsheet data to apply the practice of analyzing and interpreting data. Other practices, such as use of mathematics and computational thinking, fit naturally within this practice, because mathematical representations of relationships associated with a particular phenomenon allow students to make sense of causality and effect.

When I applied backward design to the first three stages of planning, I began with the engage phase of the 5E sequence. On day 1 of the unit, students viewed a NASA satellite image of Lake Erie. In groups, they developed questions based on observations. Typical questions included, "What is all that green stuff in the lake between Cleveland and Toledo?" and "Does all the green stuff affect the fish?" Next, students viewed an image of a Lake Erie beach showing an algal bloom and a sign prohibiting swimming. This second photograph resonated with students because the community obtains its drinking water from Lake Erie. The image made the algal bloom phenomenon more relevant to students by building on common experiences like playing on the shore, relaxing on the beach, and swimming in the water. Because the community cares about these findings, all students became invested in the learning experience. Selecting phenomena that affect all members of the community is one way to address equity in the classroom.

On day 2, the explore phase of the instructional sequence, students defined the system under study. They considered energy transfers into, out of, and within the ecosystem and ways these factors might contribute to the algal bloom phenomenon. During explorations, students attempted to answer questions from the engage phase of the sequence. However, once the system had been defined, students found themselves creating additional questions. In this phase, time was allocated for students to talk through their emerging ideas both collectively within groups and as a whole class. The opportunity to articulate ideas and begin to reason provides an impetus for students to reflect on aspects of the phenomena they do and do not understand (NRC 2008). The explore phase also enables students to consider and process various conceptions as classmates describe relevant terms and concepts.

The next day, I provided students with links to articles about causes of algal blooms in Lake Erie. We began class by reviewing those articles. Based on the readings, students then suggested possible explanations for the phenomenon. I began the explanation phase by introducing scientific ideas necessary for understanding algal blooms, such as the difference between point source and nonpoint source pollution. Additionally, I provided instruction on how to use a Secchi disk to determine the turbidity of the water in our classroom aquarium. Afterward, I supplied students with Lake Erie algal bloom data in the form of a spreadsheet. Ideally, of course, students would have traveled to the lake and collected data themselves, but time and monetary constraints make this impractical. However, analyzing the same type of data within the classroom

equipped students with the knowledge necessary to debate possible implications of the data. Using the key on the spreadsheet, students identified parameter abbreviations (e.g., *TSS* for *total suspended solids*) and talked about ideas for addressing outlier data. The introduction of relevant scientific ideas brought the science practice of analyzing and interpreting data to the foreground during this instructional sequence.

During the next class, we began the elaboration phase by analyzing the spreadsheet data and determining patterns. Students were directed to note positive correlations between total phosphorus (Total P) and total suspended solids (TSS) and negative correlations between Secchi disk depth (m) and total suspended solids (mg per m). During this phase, students integrated science and mathematics by determining the mean for Total P and TSS for each set of dates on the spreadsheet.

Students were also required to arrange data points numerically and calculate median and range. To keep practices in the foreground, we conducted a brief class discussion on analyzing and interpreting data. Applying data analysis to the learning, students worked in groups to plot graphs that conveyed the meaning of the data. Transferring mathematical graphing skills to science classrooms requires guidance, so it was sometimes necessary for me to remind students to think about scales on the axes and correctly labeling the key. Although students seemed to know the mechanics of plotting data points, they were often less proficient at interpreting the graph's meaning. For this reason, I found it beneficial to have students share their work and interpret information presented by others after preparing their own graphs.

When sharing ideas, students become actively engaged in science and engineering practices by making and defending statements about their understandings, and they are provided occasions for examining their own reasoning and conceptualizations (NRC 2005). Using student-made graphs, the practice of engaging in argument from evidence is brought to the foreground. As we focused on the positive correlation between Total P and TSS, I asked students to consider the best way to remedy the algal bloom problem. Using the graph as evidence, students engaged in argumentation about the roles Total P and TSS played in the phenomenon. Keeley (2008) notes that considering different points of view helps reinforce the value that science places on examining alternative ideas. When students engage in true argumentation, they must justify an idea and apply evidence and reasoning to support their viewpoints. Discourse via classroom discussion and argumentation allows students to enhance conceptual understanding and strengthen their scientific reasoning capabilities (Huff and Bybee 2013). The use of persuasive language also promotes critical thinking while developing students' scientific knowledge.

I kept the following three important points in mind during the elaboration phase of the instructional sequence:

1. The ability to manipulate numbers is not equivalent to the ability to analyze and interpret data. Superficial review of columns and rows on a spreadsheet is not enough to derive meaning. *Science Teachers' Learning* (NASEM 2015) asserts that there is an American tendency to teach science facts without checking for understanding. Requiring students to engage in science practices and identify relationships associated with the algal bloom phenomenon (rather than simply providing information and expecting students to recall it) allows students to develop a deeper and broader understanding of core ideas.

2. To derive meaning from data, students must communicate their ideas and the reasons behind them, and they must use those ideas to construct an explanation of the phenomenon. Before constructing their graphs, students should be provided an opportunity to discuss which variables are independent and which are dependent, what labels are most appropriate for the *x*- and *y*- axes, and what intervals to use for each axis. Moulding, Bybee, and Paulson (2015) maintain that student performances based in communication, such as speaking and writing, provide opportunities for a natural integration with productive modes of literacy. In the algal bloom example, skills emphasized included constructing an explanation and determining correlations. (The scientific practice of constructing an explanation is in the foreground because it is a hallmark of science and an essential outcome of science education.)

3. A shift in the storyline occurs in this phase. At the onset of the sequence, the storyline focuses on life science concepts pertaining to ecosystems. Within this elaboration phase, the sequence shifts into Earth and space science and the effects of humans on the environment. Krajcik et al. (2014) note that progressing through lessons that bundle performance expectations helps students to build proficiency by analyzing and interpreting data. Revisiting the storyline at this juncture also serves as a powerful reflection tool, allowing students to think about how their understanding of core ideas and ability to engage in science practices have improved over time.

The last series of lessons in the instructional sequence centers on evaluation. During this phase, students presented completed graphs and gave explanations for patterns in the ecosystem that they supported with evidence from data. Prior to writing paragraphs that explained the correlational relationships of the algal bloom, students completed a writing scaffold that helped them to make their thinking visible. The scaffold included the following prompts: *Correlations in the algal bloom data include* ___, *Changes to the system are caused by* ___, and *The phenomenon occurring in the Lake Erie system indicates* ___ *which causes* ___ *to change*. Students were required to supply evidence to support their answers.

As a teacher, I also reflected on the storyline at the conclusion of the instructional sequence. My personal reflection questions included the following:

1. Did students have sufficient opportunities to develop an understanding of core ideas, crosscutting concepts, and science practices?

2. Was instruction aligned with formative assessments during the sequence?

3. Were students provided opportunities during the learning to make connections to the *Common Core State Standards* in English language arts and mathematics?

At the conclusion of the sequence, I also examined *NGSS* evidence statements and reflected on alignment and fidelity of this unit to the standards.

Conclusion

In this particular investigation of the Lake Erie algal bloom phenomenon, science knowledge and skills are made relevant and meaningful to students because core ideas and crosscutting concepts are connected to an issue with personal meaning for students. This motivates students to engage in the investigation, which they do by constructing explanations and predicting outcomes. In this way, students discover the usefulness of science in their lives by applying science practices to make sense of the natural world.

Research on the teaching and learning of science and engineering inspired the *Framework* and the *NGSS*. Deep understanding of subject matter transforms disjointed, discrete facts into usable knowledge (NRC 2005). The *Framework's* vision promotes an expectation that learning experiences require both knowledge and practice. Science and engineering practices establish, extend, and refine knowledge.

Knowledgeable individuals are more likely to be proficient at transferring information and skills into solutions for novel problems (NRC 2000). The vision of the *Framework* and the *NGSS* centers on student engagement in exercises that allow them to make sense of phenomena rather than ones that are devoid of evidence and reasoning despite being hands-on. Though it is well worth shifting students from a focus on knowing discrete facts and memorized information to truly engaging in practices, doing so requires educators to devote considerable time, energy, and resources to the task.

A Personal Reflection on Keys to Unlocking the Vision of the *Framework* and the *NGSS*

I believe that we need to redefine teaching with a sharp focus on cultivating the habits of mind students need to negotiate a path through an increasingly complex society. Educational research tells us that today's classrooms should focus on exercises that promote the joy of learning and that encourage students to think critically. Teaching should focus on providing a deeper understanding as well as transferable knowledge of content (NRC 2012b). Twentieth-century Gestalt psychologists referred to this approach as meaningful learning. As the teaching profession adapts to promote a better understanding of the deeper structure of problems and the methods used to solve them, students will be better able to transfer their knowledge and skills to new problems and fields of study. This stands in stark contrast to more traditional lesson formats that focus on rote memorization, because the ability to simply recall facts or follow procedure does not necessarily result in the ability to transfer knowledge to related concepts or problems.

When students engage with phenomena through use of science and engineering practices and share ideas respectfully with their classmates, they learn from one another as scientists do. When students disagree on conclusions, they will devise new ways to test ideas, again modeling a scientific community. Through this process of deeper learning, students develop 21st century competencies encompassing knowledge and practices that can be transferred to new situations they encounter in science disciplines and in the workforce.

Acknowledgment

Thank you to Dwight Sieggreen for his insights in the development of this chapter.

References

Alberts, B. 2010. Reframing science standards. *Science* 329 (491): 4.

Bybee, R.W. 2013. *Translating the* NGSS *for classroom instruction.* Arlington, VA: NSTA Press.

Huff, K. L., and R. W. Bybee 2013. The practice of critical discourse in science classrooms. *Science Scope* 36 (9): 29–34.

Keeley, P. 2008. *Science formative assessment: 75 practical strategies for linking assessment, instruction, and learning.* Thousand Oaks, CA: Corwin.

Krajcik, J., S. Codere, C. Dahsah, R. Bayer, and K. Mun. 2014. Planning instruction to meet the intent of the Next Generation Science Standards. *Journal of Science Teacher Education* 25 (2): 157–175.

Moulding, B. D, R. W. Bybee, and N. Paulson. 2015. *A vision and plan for science teaching and learning: An educator's guide to* A Framework for K–12 Science Education, Next Generation Science Standards, *and state science standards.* Salt Lake City: Essential Teaching and Learning Publications.

National Academies of Sciences, Engineering, and Medicine. 2015. *Science teachers' learning: Enhancing opportunities, creating supportive contexts.* Washington, DC: National Academies Press.

National Research Council (NRC). 2000. *How people learn: Brain, mind, experience, and school.* Washington, DC: National Academies Press.

National Research Council (NRC). 2005. *How students learn: Science in the classroom.* Washington, DC: National Academies Press.

National Research Council (NRC). 2007. *Taking science to school: Learning and teaching science in grades K–8.* Washington, DC: National Academies Press.

National Research Council (NRC). 2008. *Ready, set, science! Putting research to work in K–8 science classrooms.* Washington, DC: National Academies Press.

National Research Council (NRC). 2012a. *A framework for K–12 science education: Practices, crosscutting concepts, and core ideas.* Washington, DC: National Academies Press.

National Research Council (NRC). 2012b. *Education for life and work: Developing transferable knowledge and skills in the 21st century.* Washington, DC: National Academies Press.

NGSS Lead States. 2013. *Next Generation Science Standards: For states, by states.* Washington, DC: National Academies Press. *www.nextgenscience.org/next-generation-science-standards.*

Schwab, J. 1962. The teaching of science as enquiry. In *The teaching of science,* ed. J. J. Schwab and P. F. Brandwein, 1–103. New York: Simon and Schuster.

Wiggins, G. P., and J. McTighe. 2005. *Understanding by design.* Alexandria, VA: ASCD.

Yager, R. E. 2006. *Exemplary science in Grades 5–8: Standards-based success stories.* Arlington, VA: NSTA Press.

Resources

- New York State Teaching Standards: *www.engageny.org/resource/new-york-state-teaching-standards*

- Research and Practice Collaboratory: *www.researchandpractice.org/resource/anchor_phenomena*

- *NGSS* Evidence Statements: *www.nextgenscience.org/resources/evidence-statements*

Kenneth L. Huff *is a middle school teacher in the Williamsville Central School District, Williamsville, New York. He founded and leads a Young Astronaut Council for fifth- through eighth-grade students at his school. In addition to his teaching responsibilities, Kenneth is a member of the NSTA Board of Directors and serves as the division director for middle-level science teaching; cochairs the National Academies of Sciences, Engineering, and Medicine Teacher Advisory Council; and serves as vice president of the Science Teachers Association of New York State. Huff is a past president of the Association of Presidential Awardees in Science Teaching and has served on the writing team for the NGSS. He can be contacted via e-mail at khuff@williamsvillek12.org.*

CHAPTER 2

Constructing Explanatory Arguments Based on Evidence Gathered While Investigating Natural Phenomena in the Physical Sciences in an Elementary Context

Zong Vang, Eric Brunsell, and Elizabeth Alderton

In this chapter, we address the inclusion of scientific argumentation in elementary classrooms and its connection to the *Next Generation Science Standards* (*NGSS*; NGSS Lead States 2013). We begin by discussing the importance of scientific argumentation and describing a powerful framework for constructing explanations and engaging in argumentation. We then introduce an instructional sequence for embedding argumentation into science teaching by activating prior knowledge to generate an initial explanation, building content knowledge, and critiquing and revising explanations. Finally, we identify practical strategies that can be used with elementary students to introduce, practice, and support argumentative writing and discourse.

Reinvigorated from recess, Mrs. Vang's second-grade students streamed back into the classroom. They became even more excited when a police officer walked into the room. "Are we in trouble, or are we learning more about communities today?" joked one student. Mrs. Vang reminded students that they had been talking about police and other people in the community yesterday during social studies, and she thought it would be fun to invite her husband, a detective for the police department, to visit the class.

After introducing himself, Detective Vang asked the students what they thought he did as a detective. "You arrest bad guys," said one student. After a few more responses, another student said, "You solve crimes." Detective Vang then gave a short presentation on how he solved crimes:

making observations, collecting evidence, and then trying to figure out what that evidence means. He showed them his camera, how he takes a fingerprint off of a mirror, and other tools of the trade. Then, he explained how he writes a report that includes his "claim," or what he thinks happened, which is supported by "evidence" and "reasoning," which is an explanation of how the evidence supports his claim. Mrs. Vang wrote the words claims, evidence, *and* reasoning *on the board. Later, she would use those terms and definitions to create an anchor chart that she can use across all subjects as students share their ideas and thought processes.*

After Detective Vang left, Mrs. Vang asked the students to list ways that detectives and scientists are similar. "They figure stuff out," exclaimed one student. Mrs. Vang explained that scientists also use evidence to make claims about how the world works. "Let's try it," she said, while handing out bowls of fossil fragments. "Use your evidence—your observations—to make a claim about what these things are."

Arguing in Science

In his workshops, Eric Brunsell often starts by asking teachers to use one phrase to define what scientists do. Participants shout out words and phrases: *discovery, research, collect data, explain the world,* and so on. He summarizes these words into one sentence: "Scientists use evidence to explain the natural world"—or put even more simply, "Scientists argue." This is similar to Martin and Hand's assertion that "Argumentation is a central tenet of science" (2009, p. 20) and the American Association for the Advancement of Science's (1993) definition of science inquiry as the use of evidence and logic to explain the natural world.

If argumentation is central to science, then it should also be central to classroom activities—a conclusion solidly supported by research (e.g., Driver, Newton, and Osborne 2000; NRC 2007; NRC 2012) and encoded in the *NGSS*'s emphasis on using evidence to create explanations, engaging in argumentative discourse, and evaluating information from multiple sources. Rubrics that show the grade-band expectations for science and engineering practices most directly related to argumentation are included in Tables 2.1 (p. 13) and 2.2 (p. 14).

McNeill and Krajcik (2011) provide a powerful model for constructing explanations and engaging in argumentation with their claim, evidence, and reasoning (CER) framework. This model works well in science as well as in other subjects (e.g., for instilling such argumentation-based English language arts skills as persuasive writing). We use an anchor chart in our classroom that spells out the components of the CER framework as follows:

- **Claim:** The claim is a statement that answers the original question. A good claim should be testable.

- **Evidence:** Evidence is the scientific data that support a claim. It can come from observations, investigations, or from secondary source material (e.g., a reading).

- **Reasoning:** Reasoning is the justification that explains how the evidence supports the claim.

In our classrooms, we explicitly teach argumentation through reading, writing, speaking, and listening, allowing us to build connections with related skills in social studies, language arts, and mathematics. In this chapter, we will explore example lessons that place argumentation at the center of instruction, describe strategies for introducing argumentation to elementary students, and introduce scaffolds for supporting students' ability to engage in argumentative discourse.

Including Argumentation in Instruction

Science educators have known for decades that students learn better by first engaging with science concepts in a hands-on manner and then making sense of those activities in order to build formal content understanding (Abraham and Renner 1986; Karplus and Thier 1967; Roth 1989). Sampson, Grooms, and Walker (2009) have designed and successfully implemented a cyclical approach to building scientific understanding through argumentation. The Argument Driven Inquiry model begins with a focus task that requires students to either make sense of a phenomenon or solve a problem. Students then collect data and craft a tentative argument in small groups. Next, the entire class shares and critiques those arguments. Groups or individuals then revise their argument, participate in a peer-review process, and develop a "final" argument. Although this model was initially used with middle and high school students, Chen et al. (2016) found that a modified version also had positive effects when used with younger students.

We use a simplified argumentation model that incorporates elements of a more traditional learning cycle approach. This model includes the following components:

- **Activating** prior knowledge by providing students with a task that allows them to create initial arguments for a phenomenon that they share and critique with one another

- **Building** content knowledge through investigations, readings, and other activities

- **Critiquing** and revising arguments so that they are more scientifically appropriate

The acronym here, *ABC,* reinforces the **A**ctivity **B**efore **C**ontent approach at the core of any effective

Table 2.1. *NGSS* Science and Engineering Practices Grade Band Endpoints for Grades K–2

1	2	3	4				
The student shows no evidence of being able to engage in this practice.	The student can engage in this practice with help.	The student can engage in this practice by him or herself.	The student excels at this practice.				
Constructing explanations and designing solutions in K–2 builds on prior experiences and progresses to the use of evidence and ideas in constructing evidence-based accounts of natural phenomena and designing solutions.							
Make observations (firsthand or from media) to construct an evidence-based account for natural phenomena.				1	2	3	4
Use tools and/or materials to design and/or build a device that solves a specific problem or a solution to a specific problem.				1	2	3	4
Generate and/or compare multiple solutions to a problem.				1	2	3	4
Engaging in argument from evidence in K–2 builds on prior experiences and progresses to comparing ideas and representations about the natural and designed world(s).							
Identify arguments that are supported by evidence.				1	2	3	4
Distinguish between explanations that account for all gathered evidence and those that do not.				1	2	3	4
Analyze why some evidence is relevant to a scientific question and some is not.				1	2	3	4
Listen actively to arguments to indicate agreement or disagreement based on evidence, and/or to retell the main points of the argument.				1	2	3	4
Construct an argument with evidence to support a claim.				1	2	3	4
Make a claim about the effectiveness of an object, a tool, or a solution that is supported by relevant evidence.				1	2	3	4
Obtaining, evaluating, and communicating information in K–2 builds on prior experiences and uses observations and texts to communicate new information.							
Read grade-appropriate texts and/or use media to obtain scientific and/or technical information to determine patterns in and/or evidence about the natural and designed world(s).				1	2	3	4
Describe how specific images (e.g., a diagram showing how a machine works) support a scientific or engineering idea.				1	2	3	4
Obtain information using various texts, text features (e.g., headings, tables of contents, glossaries, electronic menus, icons) and other media that will be useful in answering a scientific question and/or supporting a scientific claim.				1	2	3	4
Communicate information or design ideas and/or solutions with oral and/or written forms using models, drawings, writings, or numbers that provide detail about scientific ideas, practices, and/or design ideas.				1	2	3	4

Source: Adapted from Brunsell, Kneser, and Niemi 2014.

Table 2.2. *NGSS* Science and Engineering Practices Grade Band Endpoints for Grades 3–5

1	2	3	4				
The student shows no evidence of being able to engage in this practice.	The student can engage in this practice with help.	The student can engage in this practice by him or herself.	The student excels at this practice.				
Constructing explanations and designing solutions in 3–5 builds on K–2 experiences and progresses to the use of evidence in constructing explanations that specify variables that describe and predict phenomena and in designing multiple solutions to design problems.							
Construct an explanation of observed relationships (e.g., the distribution of plants in the backyard).				1	2	3	4
Use evidence (e.g., measurements, observations, patterns) to construct or support an explanation or design a solution to a problem.				1	2	3	4
Identify the evidence that supports particular points in an explanation.				1	2	3	4
Apply scientific ideas to solve design problems.				1	2	3	4
Generate and compute multiple solutions to a problem based on how well they meet criteria and constraints of the design solution.				1	2	3	4
Engaging in argument from evidence in 3–5 builds on K–2 experiences and progresses to critiquing the scientific explanations or solutions proposed by peers by citing relevant evidence about the natural and designed world(s).							
Compare and refine arguments based on an evaluation of the evidence presented.				1	2	3	4
Distinguish among facts, reasoned judgment based on research findings, and speculation in an explanation.				1	2	3	4
Respectfully provide and receive critiques from peers about a proposed procedure, explanation, or model by citing relevant evidence and posing specific questions.				1	2	3	4
Construct and/or support an argument with evidence, data, and/or a model.				1	2	3	4
Use data to evaluate claims about cause and effect.				1	2	3	4
Make the claim about the merit of a solution to a problem by citing relevant evidence about how it meets the criteria and constraints of a problem.				1	2	3	4
Obtaining, evaluating, and communicating information in 3–5 builds on K–2 experiences and progresses to evaluating the merit and accuracy of ideas and methods.							
Read and comprehend grade-appropriate complex texts and/or other reliable media to summarize and obtain scientific and technical ideas and describe how they are supported by evidence.				1	2	3	4
Compare and/or combine across complex texts and/or other reliable media to support the engagement in other scientific and/or engineering practices.				1	2	3	4
Obtain and combine information from books and/or other reliable media to explain phenomena or solutions to a design problem.				1	2	3	4
Communicate scientific and/or technical information orally and/or in written formats, including various forms of media and may include tables, diagrams, and charts				1	2	3	4

Source: Adapted from Brunsell, Kneser, and Niemi 2014.

learning cycle. Our science units are divided into multiple cycles, and in each one we try to focus on only one or two major concepts. Tables 2.3 and 2.4 provide examples that illustrate the ABC cycle in the elementary classroom.

Introducing Argumentation to Students

We introduce argumentative discourse to students at the beginning of the school year in a manner that allows them to connect their everyday speech to academic language (Varelas, Pappas, Kane, and Arsenawlt 2008). The opening vignette in this chapter is an excellent example, bridging students' understanding of detective work with the academic process of argumentation or CER. Hong et al. (2013) provide another example of using everyday experiences to introduce inquiry. In their classroom, they have students investigate ways to thicken a soup and then present their arguments to the rest of class. In our classroom, we often use realia (e.g., fossils) or close-up images of common objects (e.g., a cat's tongue) to reinforce argumentation. Our students make observations that they then use as evidence to support a claim that identifies the object. We have also used mystery theater (e.g., following clues to determine what happened to a toy or the cookie jar) to act out crime scenes and help students understand the use of evidence to support a claim.

Once students have a basic understanding of the CER framework, they can begin to identify arguments in readings. Teaching students to identify arguments in text prior to having them write their own arguments is a core element of persuasive (or opinion) writing instruction, and this technique should be applied across the disciplines. For example, when we read the book *Duck! Rabbit!* by Rosenthal and Lichtenfield (2009) with our students, we have them identify the evidence that the two main characters use to convince each other that an

Table 2.3. First-Grade Instructional Sequence Using the ABC Cycle

Science Concept: Sound is caused when an object vibrates (PS4.A)	
Activate Prior Knowledge	Students complete an engineering design challenge in which they attempt to create a better cup-and-string phone. After experiencing the cup-and-string phone, students are asked to create an annotated illustration that explains how sound gets from one cup to the other.
Build Content Knowledge	Students investigate a variety of noisemakers (e.g., bells, musical birthday cards, their vocal cords, tuning forks) and record observations about what happens when sound is made. A class discussion concludes that all of these objects vibrate. The teacher reads the book *What Makes Different Sounds* (Lowery 2013) to further reinforce the concept that sound is created by vibrations.
Revise and Critique	Students use their new knowledge to revise their initial annotated illustrations for how sound gets from one cup to the other. The teacher specifically asks students to include evidence as to how they know that vibrations cause sound.

Table 2.4. Fourth-Grade Instructional Sequence Using the ABC Cycle

Science Concept: Energy can transform from potential to kinetic energy in a moving object (PS3.A)	
Activate Prior Knowledge	Students are presented with the claim, "The first hill of a roller coaster must be the highest point of a roller coaster." Students are then challenged to create roller coasters with multiple hills and loops. To be successful, a marble must remain on the track for the entire coaster. After students observe coasters created by other groups, they support the provided claim with evidence and reasoning.
Build Content Knowledge	Students read their textbook, watch a video, and discuss potential and kinetic energy.
Revise and Critique	Students revise their reasoning in the initial roller coaster argument to better represent their more scientific understanding of potential and kinetic energy.

animal is either a duck or a rabbit. Early elementary students can easily identify the competing claims made in the book and identify the evidence used to support each claim.

As students gain more experience identifying arguments, they can do the same with science nonfiction. For example, upper elementary students can use the *Time for Kids* article "Science Turns Your World Gray" (Basu 2015) to identify the evidence and claims presented about the ways in which mood can affect perception.

Scaffolding to Engage Students in Argumentation

We frequently use graphic organizers and peer review to help students engage in argumentative writing and discussions, and we reinforce argumentative discourse by providing students with specific discussion strategies. Dowell, Tscholl, Gladisch, and Asgari-Targhi (2009) found that graphic organizers improve student arguments by helping them to visually represent the components of an argument.

We have used a variety of graphic organizers to support argumentation in elementary science. For example, during a second-grade unit on properties of matter, students explored the flexibility of different materials and recorded their observations in a simple table (see Figure 2.1). After finishing their observations, students engaged in a class discussion to collaboratively create an argument that identified the most flexible material. I recorded this argument on chart paper as follows:

- **Claim:** *I think* ... the yarn is the most flexible.
- **Evidence:** *Because* ... the yarn moved easily. I didn't have to force it to tie it in a knot. The yarn was soft, not hard. I couldn't even bend the marker. The licorice broke when we tried to tie it in a knot.
- **Reasoning:** The yarn was made of smaller threads that were softer and smaller than the licorice and markers. The material of the yarn could bend. The material of the licorice could only bend a little, and the marker materials didn't bend at all.

Later in that same unit, the students explored the *NGSS* disciplinary core idea 2-PS1.B: "Heating or cooling a substance may cause changes that can be observed. Sometimes these changes are reversible, and sometimes they are not." The students engaged in this

Figure 2.1. Student Observations of the Flexibility of Different Materials

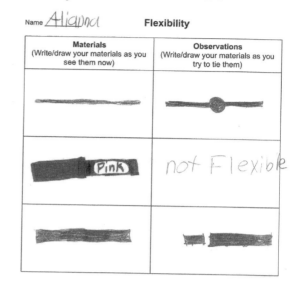

Order your materials from most flexible to least flexible:

yarn
twizzler
marker

investigation by periodically observing ice in beakers sitting on a counter. Students drew their observations in "evidence buckets," (see Figure 2.2) and I also recorded results for the class. Students then used a "Claims and Evidence" organizer that provided sentence starters to support their writing.

Crowhurst (1990) found that students who collaborated while critiquing written arguments did significantly better than students who critiqued alone. Sampson, Grooms, and Walker (2009) also encourage the use of peer review to help students engage in argumentation.

Elizabeth Alderton, a colleague at the University of Wisconsin-Oshkosh, uses a "paper blogging" approach with students to facilitate peer review. Students complete the first portion of a three-part organizer with the initial draft of their argument, then trade papers and use the second part of the form to give feedback on a peer's argument. Finally, students get their papers back and can use that feedback to revise their initial argument.

Figure 2.2. Example of a Student's Evidence Buckets to Record Evidence of the Melting of Ice

During this process, students are coached to critique an argument by answering the following questions:

- Does the argument include a claim?
- Does the author use evidence to support the claim?
- Does the author show his or her reasoning by using a science idea?
- What could the author do to make the writing better?

Lee and Kinzie (2012) found that student engagement in argumentation is supported when teachers use open-ended questions during discussions that specifically target the higher-order cognitive skills they wish to reinforce. The approaches that we use for argumentation in science are similar to those we use to make thought processes visible in mathematics. The *Common Core State Standards for Mathematics* standard MP3—"Construct viable arguments and critique the reasoning of others"—asks students to explain the process they used to arrive at their answers and also to speculate about different processes that others may have used to arrive at the same answer. In both math and science, teachers can foster deep discussions by using questions like the following:

- Can you show me why?
- Can you prove it to me? What is your evidence?
- Do you agree with this? Why or why not?
- Why do you think this answer is incorrect?

- Where do you think this person went wrong, based on his or her evidence?

Students should also be explicitly taught to constructively challenge each other during a discussion by working as a class to set discussion norms, providing students with specific discussion moves, and modeling expected behavior. I hang an anchor chart titled "Challenge Others' Thinking" on the wall with the following moves that students can use during discussion:

- Can you show me?
- Explain how …
- How do you know?
- What is your evidence?
- Why do you think that …?
- Tell me more …
- Can you give me an example?
- Prove it!
- I disagree with you, because …

I also use the following discussion norms: wait for someone to finish talking first; disagree with the statement, not the person—be friendly; and talk to your classmates, not to your teacher.

Adding the Crosscutting Concepts to Instruction

The *NGSS* crosscutting concepts represent underlying themes that are important across all of the disciplines of science. According to the *Framework,* they "provide a connective structure that supports students' understanding of sciences as disciplines and that facilitates their comprehension of the system under study in particular disciplines" (pp. 4–13). In this sense, they should be used along with the science and engineering practices and the core ideas as thinking tools to help students develop an understanding of science. When students encounter a new phenomenon, they should begin asking themselves questions to help them begin to explain what they are seeing. For example:

- *Patterns:* What patterns do I notice? Have I seen patterns like this before?
- *Cause and Effect:* What might cause the effect that I am seeing?
- *Scale, Proportion, and Quantity:* Does the amount of 'stuff' matter?
- *Systems and System Models:* What is the system that we are observing? How do the parts of that system work together?
- *Energy and Matter:* What is happening to the matter or energy? (*Note:* Energy is not introduced until grade 4.)
- *Structure and Function:* How are the structure and function related?
- *Stability and Change:* What is changing? What is staying the same?

Conclusion

The *Framework* and the *NGSS* present a vision that integrates scientific explanations with the practices of science and engineering that are needed to generate them. The process of argumentation, defined as the use of evidence to construct and revise explanations of phenomena, is central to this vision. Our experiences have shown us that argumentation instruction not only engages but also motivates young children to think critically. Students are continually questioning and collaborating with one another to develop deep conversations about every subject, not just science. In math, they ask classmates what they did to get their answer and they show one another the processes that they used. In reading, students infer and are asked by other book club members what evidence they have gathered from the book to support their thinking. We

have also learned that by setting argumentation norms, students understand and accept disagreements. They feel comfortable and safe with their classmates, even while being questioned and challenged. In a classroom that supports argumentation, students understand that being challenged by their peers helps them to become better learners, thinkers, and scientists. Furthermore, by teaching students to argue effectively and strive for deeper understanding, we are facilitating the development of 21st-century college and career readiness. Our experiences implementing the vision of the *NGSS* into an elementary school classroom has shown that placing argumentation at the center of instruction positively affects students' learning and enjoyment, not only of science but of all disciplines that benefit from critical thinking and effective peer-to-peer communication.

Additional Resources

The following books can be used to reinforce argumentation in an elementary classroom:

- *A Fine, Fine, School* by Sharon Creech and Harry Bliss (HarperCollins, 2003)
- *Earrings!* by Judith Viorst (Atheneum Books for Young Readers,1993)
- *I Wanna Iguana* by Karen Kaufman Orloff and *I Wanna New Room* (G.P. Putnam's Sons Books for Young Readers, 2004 and 2010)
- *My Teacher for President* by Kay Winters and Denise Brunkus (Puffin Books, 2008)
- *The Perfect Pet* by Margie Palatini (Katherine Tegen Books, 2009)
- *The Spider and the Fly* by Mary Howitt and Tony DiTerlizzi (Simon & Schuster Books for Young Readers, 2002)

References

Abraham, M. R., and J. W. Renner. 1986. The sequence of learning cycle activities in high school chemistry. *Journal of Research in Science Teaching* 23 (2): 121–143.

American Association for the Advancement of Science. 1993. *Benchmarks for science literacy.* New York: Oxford University Press.

Basu, T. 2015. Sadness turns your world gray. *Time for Kids. www.timeforkids.com/news/ sadness-turns-your-world-gray/269151.*

Brunsell E., D. Kneser, and K. Niemi. 2014. *Introducing teachers and administrators to the* NGSS. Arlington, VA: NSTA Press.

Chen, H. T., H. H. Wang, Y. Y. Lu, H. S. Lin, Z. R. Hong. 2016. Using a modified argument-driven inquiry to promote elementary school students' engagement in learning science and argumentation. *International Journal of Science Education* 38 (2): 170–191.

Crowhurst, M. 1990. Teaching and learning the writing of persuasive/argumentative discourse. *Canadian Journal of Education* 15 (4): 348–359.

Dowell, J., M. Tscholl, T. Gladisch, and M. Asgari-Targhi. 2009. Argumentation scheme and shared online diagramming in case-based collaborative learning. *Computer-supported collaborative learning practices: CSCL 2009 conference proceedings.* Vol. 1, eds. C. O'Malley, D. Suthers, P. Reimann, and A. Dimitracopoulou, 567–575. Rhodes, Greece: International Society of the Learning Sciences.

Driver, R., P. Newton, and J. Osborne. 2000. Establishing the norms of scientific argumentation in classrooms. *Science Education* 84 (30): 287–312.

Hong Z. R., H. S. Lin, H.H Wang, H.T. Chen, and K. K. Yang. 2013. Promoting and scaffolding elementary school students' attitudes toward science and argumentation through a science and society intervention. *International Journal of Science Education* 35 (10): 1625–1648.

Karplus, R., and H. D. Thier. 1967. *A new look at elementary school science.* Chicago: Rand McNally.

Lee, Y., and M. Kinzie. 2012. Teacher question and student response with regard to cognition and language use. *Instructional Science* 40 (6): 857–874.

Lowery, L. F. 2013. *What makes different sounds?* Arlington, VA: NSTA Press.

Martin, A. M., and B. Hand. 2009. Factors affecting the implementation of argument in the elementary science classroom: A longitudinal case study. *Research in Science Education* 39 (1): 17–38.

McNeill, K. L., and J. Krajcik. 2011. *Supporting grade 5–8 students in constructing explanations in science: The claim, evidence and reasoning framework for talk and writing.* New York: Pearson Allyn & Bacon.

National Research Council (NRC). 2007. *Taking science to school: Learning and teaching science in grades K–8.* Washington, DC: National Academies Press.

National Research Council (NRC). 2012. *A framework for K–12 science education: Practices, crosscutting concepts, and core ideas.* Washington, DC: National Academies Press.

NGSS Lead States. 2013. *Next Generation Science Standards: For states, by states.* Washington, DC: National Academies Press. *www.nextgenscience.org/next-generation-science-standards.*

Rosenthal, K., and T. Lichtenfield. 2009. *Duck! Rabbit!* San Francisco: Chronicle Books.

Roth, K. J. 1989. Science education: It's not enough to "do" or "relate." *American Educator* 13 (4): 46–48.

Sampson, V., J. Grooms, and J. Walker. 2009. Argument-driven inquiry. *The Science Teacher* 76 (8): 42–47.

Varelas, M., C. C. Pappas, J. M. Kane, and A. Arsenawlt. 2008. Urban primary-grade children think and talk science: Curricular and instructional practices that nurture participation and argumentation. *Science Education* 92: 65–95.

Zong Vang *is a second-grade teacher at Webster Stanley Elementary School in Oshkosh, Wisconsin. She received her MS in education from the University of Wisconsin-Oshkosh and is also on the Science Task Force team for the Oshkosh Area School District.*

Eric Brunsell *is associate professor of science education and the interim director of professional education programs for the University of Wisconsin-Oshkosh. He is also the chief operations officer for the Wisconsin Society of Science Teachers and serves on NSTA's Board of Directors. Brunsell can be reached by e-mail at brunsele@uwosh.edu.*

Elizabeth Alderton *is associate professor of literacy and assistant dean of the College of Education and Human Services for the University of Wisconsin-Oshkosh. She is actively involved in PreK–12 literacy initiatives and serves on the Board for the Educational Foundation of Neenah in Wisconsin.*

CHAPTER 3

Engaging Students in Disciplinary Core Ideas Through the Integration of Science and Engineering Practices While Making Connections to the Crosscutting Concepts

Mary Colson

In this chapter, we explore changes to instructional planning that support three-dimensional learning, including specific techniques for helping students figure out possible explanations for observable phenomena and engage in robust discourse and argumentation. We examine the use of storylines to organize lessons around phenomena and investigative questions. To conclude the chapter, we discuss teachers as *practitioners of science*, actively engaged in the practices of science during instruction.

Introduction: Now and Then

Several years ago, just before the rollout of the *Next Generation Science Standards* (*NGSS*; NGSS Lead States 2013), my friend and colleague Jean Tushie asked me, "With all your work on the *NGSS* writing team, what are you doing now that is different from how you used to teach?" As I recollect, I didn't have a particularly good answer—something like "I already do everything just right." This was an inadequate response to an important question. During our careers, Jean and I have seen many movements in education come and go without having much of an effect on instruction, and she wanted to know if the *NGSS* had the potential to truly make a difference.

Now, as I have worked to enact the vision of the *NGSS*, my answer to her question would be that I've changed my instruction in significant ways. For 24 years, I thought that my students were actively engaged in "doing science," but I now realize that this was often

not the case. Instead, my students were actively engaged merely in confirming what they had been told. To support three-dimensional learning as envisioned by the *NGSS*, I have shifted my role in the classroom so that I now act more as a facilitator who empowers students to make sense of the world through the practices of science. I now strive to

- organize my curriculum around investigative questions about phenomena rather than topics or chapter headings,

- structure opportunities for scientific inquiry that allow my students and I to figure stuff out like scientists rather than to confirm or reinforce ideas, and

- facilitate and moderate student-to-student discourse discussions.

Organizing Lessons and Units Around Investigative Questions and Phenomena

I have dozens of digital and paper folders full of notes, lessons, and student work. They are organized by topic, and the labels read like the table of contents of a textbook. According to *A Framework for K–12 Science Education* (*Framework*; NRC 2012), the *NGSS* disciplinary core ideas, however, are not intended to be an exhaustive list of science concepts but rather a coherent set of explanatory ideas. These ideas have broad importance across and within the science disciplines and "provide a key tool for understanding or investigating more complex ideas and solving problems" (p. 31). Engaging students in the disciplinary core ideas means that students have repeated opportunities to question, observe, and investigate much in the same manner as a scientist. They apply those observations and their understanding to the core ideas and, over time, their understanding deepens. When asked what they're doing in science class, I want my students to be able to answer, "We are trying to figure out *X*" rather than "We are studying *X* because the teacher said to."

The *Framework* emphasizes the importance of questions in science education: "Each core idea and its components are introduced with a question designed to show some aspect of the world that this idea helps to explain. The question is followed by a description of the understanding about the idea that should be developed by the end of high school. This structure is intended to stress that posing questions about the world and seeking to answer them is fundamental to doing science" (p. 31–32). To implement this structure, I have been shifting the organization of my curriculum for the last several years to center around questions and phenomena that my students can investigate rather than discrete topics that I want them to understand. I use the metaphor of a storyline (Reiser 2014; Colson R. and M. Colson 2016) to hook consecutive lessons together in a coherent way. Like scenes in the storyline of a novel, each lesson sets the stage for the next by getting the students to think about new questions to answer.

Storylines: An Organizational Scheme for Investigative Questions and Phenomena

One storyline I've been working on for the last few years addresses standard ESS1.B: Earth and the Solar System of the *NGSS* and is anchored by the following student question: "The Moon can't be out during the day, can it?" In this series of investigations, summarized in Table 3.1, students incrementally develop a model of motion for bodies in the Earth-Moon-Sun system to explain an array of observations that we can see from Earth. In the following brief summary of this storyline, each lesson provides a reason, a question, or an observation that compels students to figure out something new.

My students start by considering the Earth-Sun system only. I task them with finding the answer to the question, "Which way is the Earth rotating?" They work in our darkened classroom, using a flashlight Sun and a tennis-ball Earth with a pin stuck in it to mark where we live. The students rotate the little globe and watch the where-we-live pin rotate into the darkness and back into the light. They work at connecting the changing location of where we live with time of day we would be experiencing. Students develop an explanatory model for how rotation can cause the day-night cycle, which helps them figure out which way the Earth is rotating. They support their claim with evidence and reasoning derived from their models (see Table 3.1, lesson-level question 1).

During the first lesson, students invariably ask, "What's the Moon doing while Earth is rotating?" I use that question to help transition to the next scene, although I rephrase it as "What can we observe from Earth about the patterns of the Moon's changing appearance?" so that we can find our answer through observation. The students then explore changes in the Moon's appearance and look for patterns in them. They develop a summary of the patterns they see and a common language for referring to the different shapes of the Moon (see Table 3.1, lesson-level question 2).

The patterns from the second lesson then become the evidence for the third "scene," during which students develop a model to explain how the Moon's orbit causes the phases we see from our Earth-based perspective. Adding a small Styrofoam Moon to their Earth-Sun system, students now consider the Earth's rotation and the moon's revolution as they develop a way to explain the pattern of lunar phase. Requiring students

Table 3.1. Summary of Storyline for Developing a Model to Explain Phases and Eclipses.

Big Idea: The Earth and Moon are part of a much larger solar system and all the objects in the solar system are in motion.			
Question from student that began the series of investigations: The Moon can't be out during the day, can it?			
Lesson-level question	**Phenomenon to explore**	**What students did**	**What students figured out**
1. Which way is the Earth rotating?	Cycle of day followed by night, over and over again	They investigated, modeled, and explained while testing and revising their own ideas, using a model of only the Sun and Earth).	Earth spins counterclockwise, as one looks at the North Pole. Earth is always half in shadow. The shadowed side is the side that doesn't face the Sun.
During the course of the first lesson, a common question that popped up was, "What's the Moon doing while we are rotating?" This question formed the basis for the next scene of our storyline.			
2. What are the patterns we can observe from Earth in the Moon's changing appearance?	Cycle of Moon's phases	They observed the different patterns and figured out a way to describe them.	Important patterns to be modeled (e.g., waxing period takes about 14 days, the lighted side gets bigger from day to day)
I helped students to link the observations they made during the second lesson to the big idea (also a disciplinary core idea) that the motion of bodies in the solar system causes phenomena that we observe from Earth.			
3. How can the Moon's orbiting cause the phases we see from our Earth-based perspective?	For about 14 days, the Moon's lighted side gets bigger, and then for about 14 days the lighted side gets smaller.	They used a three-dimensional physical model of the Earth-Moon-Sun system to account for the observed patterns and argued from evidence.	The cause of Moon phases (e.g., why the waxing gibbous Moon is out at 4 p.m.)
During the preceding lesson, students usually discover that eclipses should be occurring every month, based on their model.			
4. Why don't we have eclipses twice a month?	Model suggests monthly eclipses should occur.	They modeled the angle of Moon's orbit relative to the plane of Earth's orbit.	Information regarding rotation, revolution, plane of orbits, Earth-based and "space alien" points of view of the Earth, Moon, and Sun; times of day when eclipses occur

Note: Individual lessons are in the rows with white background.

to explain why we see certain phases at certain times of day is a test of students' models and understanding (see Table 3.1, lesson-level question 3).

The unit concludes with a fourth lesson to answer the student-generated question, "Shouldn't we be having eclipses every month during full and new Moon phases?" During this lesson, students model rotation, revolution, and orbital planes. They can use their models to explain why certain phenomena, like solar eclipses, are possible only at certain times of the day. My students can mentally shift between an Earth-based perspective of looking up into the sky and a space alien's perspective of the Sun, the relative positions of the Moon and the Earth, and the shadowed and lit sides of both. At this point in the students' storylines, their models don't account for the tilt of Earth's axis, nor have students figured out how to explain the cause of Earth's seasons. Still, the model provides a firm footing for later investigations into these and other issues. (see Table 3.1, lesson-level question 4).

The work my students did in this unit was not unlike the work of real scientists, and their learning occurred at the nexus of the three key dimensions. Students asked questions that they couldn't immediately answer, and they figured out explanations that accounted for their observations. They looked for patterns, developed and used models, constructed explanations, communicated ideas, and argued from evidence (practices). As students looked for patterns, they modeled a part of the solar system and determined possible causes (crosscutting concepts). They used the disciplinary core idea that the motion of bodies in the solar system can explain a host of observations. Like real scientific research, one investigation led naturally to more questions of interest, which in turn determined the direction of the next investigation.

Structuring Opportunities for Authentic Scientific Inquiry

For years, my students did a lab on the unequal heating of land and water (Colson, M. and R. Colson 2016). In the early years, I honed the lab procedures to prevent students from introducing variables that yielded "wrong" answers, ensuring that my students got good data that they could use to explain differences in climate. The final assessment, for example, was for students to analyze the annual temperature variations for San Francisco and St. Louis and explain why San

Francisco had such little variation in temperature from summer to winter compared to St. Louis.

To student-proof the lab procedures, I tried a variety of experimental design approaches, interpreted results that didn't at first make sense, and tested the reproducibility of results. I was engaged in scientific practices and crosscutting concepts, but my students were not. In my effort to streamline the lab and make it time-efficient, I had stripped away the messiness of scientific investigation. As a result, my students didn't have the chance to figure out whether they had correctly identified and controlled for variables, to compare their results with those of their classmates, or to interpret unexpected data.

My active-learning inquiry lab reinforced knowledge presented in the textbook and lecture. The purpose of the lab was not for students to pursue questions like "How much faster does soil heat up than water?" or "How does the unequal heating of the Earth affect regional climates?" Nor was it to develop an explanation for their experimental findings. Instead, the purpose was for them to experience a tactile and memorable way of remembering what they learned. The phenomenon of unequal heating, which a barefoot kid at the beach in July can notice, was not the intellectual force behind the lesson.

The shift from teaching science as a body of knowledge to teaching science as a way of knowing will be one of the most difficult changes to carry out while implementing the *NGSS*. One challenge is providing the amount of time students need to process observations and develop or test tentative explanations. My students often require more time than I feel I can spare to develop their understanding of knowledge through science practices and crosscutting concepts. If I give them the time necessary to engage in three-dimensional learning, I must decide what other topics not to cover. That's a tough decision in this climate of accountability through standardized testing, and I find myself wondering if it is fair for my students not to encounter all the possible topics that could appear on a state test. However, I think it is more important to ask whether our educational system really tests the most important skills and concepts, or if it is even possible to truly assess those things. For better or worse, I now give much less thought to standardized tests than I used to. Instead, I focus on the day-to-day interactions with students, during which I rely on short formative assessments to guide my instruction.

A second challenge is being able to recognize who is doing the practices in the classroom. Though scientific practices were sometimes present in my student-proofed labs (when I asked students to cite evidence and reasoning that supported important concepts, for example), I rarely asked students to construct explanations, argue from evidence, or develop scientific models with predictive capabilities. I needed to figure out a structured manner of handing over some of that intellectual work to my students.

Creating a Classroom Where Three-Dimensional Learning Flourishes

Since the publication of the *Framework*, my biggest instructional shift has been to change my role in the classroom from providing information to acting as a mentoring scientist. My own past experience has informed this shift. I engaged in the practices of science most authentically during my geology master's program at the University of Tennessee. During my research, my confidence in my ability to "do" science grew as I interacted with and learned from my thesis adviser. In our book *Learning to Read the Earth and Sky,* Russ Colson and I write: "Engaging students in doing authentic science investigations—that is, investigations in which the outcome is not preordained by the textbook and where students make real choices in what questions to pursue and real contributions to the experiments and interpretations—requires that the teacher be a practitioner of science" (Colson, R. and M. Colson 2016, p. 344). I think a lot about how to arrange my students' day-to-day work so that they are positioned to investigate questions and develop their understanding.

Instructional Moves to Support Three-Dimensional Learning: A Classroom Snapshot

During a lesson investigating how rocks can form from lava, I had my students use thymol, a compound with a low melting temperature, to observe melting and crystallization firsthand. In the following classroom snapshot, I highlight a few strategies I used to create authentic learning situations and allow curiosity to blossom into fruitful work.

My students developed our initial question—"How do rocks form from lava?"—during a reading activity related to the origins of rock. I planned the reading activity specifically for students to generate questions

that we could begin to answer experimentally. As in the BSCS 5E Instructional Model's Engage phase, students began their work with play that exhibited no real goal beyond making observations and asking questions. They were, however, required to write about what they saw, what they wondered about, and what more they might need to know in their lab journals.

During play, questions tumbled from my students. Many of their best questions of the unit emerged in this risk-free environment. Rather than answer their questions directly, I asked students to write them in their journals. In this way, they were able to develop insights that allowed them to contribute meaningfully to the upcoming class discussion. When possible, I deflected questions to which I planned to circle back as the unit unfolded, giving them the opportunity to answer them on their own as their understanding developed.

I followed this time for play with a formal class discussion. The goal of the discussion was for students to develop a sense of the breadth of ideas that we might need to consider in explaining the causes of our observations. Through the discussion, each class came to a consensus about which questions to pursue further. The discussion also served as a time to activate students' background knowledge and other personal experience. For example, one student observed that the melting thymol looked like water on snow. Many students thought that the melted thymol actually *was* water. This discussion allowed students to explore their ideas in a risk-free and meaningful way, without worrying about a grade or getting a right answer.

It takes time and effort to build and support a classroom culture that values a curious and questioning stance. It takes fortitude to corral the curiosity and questions into some semblance of coherence. And it takes a practitioner of science to understand students' inchoate ideas and find ways to develop them so that student understanding becomes increasingly sophisticated. To support three-dimensional learning, I plan opportunities for students to do the following:

- Make observations and generate questions.

- Discuss ideas, observations, connections, and explanations with one another, with me, and with the whole class.

- Observe and wonder again and again, each time armed with additional knowledge about what to look for, what to think about, and what to ignore.

- Summarize their current understanding during a teacher-moderated discussion session.

- Develop an initial model or explanation, based on relevant information from outside sources and from our summaries of understanding.

- Share models and explanations with one another.

- Argue for their explanations and models and support their arguments with evidence and reasoning.

- Revise their models and explanations.

Enhancing Discourse and Argumentation

The third significant shift in my instruction, and one to which I currently devote a lot of attention, is in learning to facilitate and moderate student-to-student discourse. Over the last several years, I've paid attention to the flow of discussions in my classes, noting who talked, who listened, and when. I realized that much of what I thought was discussion actually constituted sequential mini-dialogues between myself and an individual student, with most of the other students not really listening. When I questioned students later about key ideas discussed in these mini-dialogues, more often than not they couldn't remember them.

Retooling the teacher-asks-question and student-responds-to-teacher habit is difficult. It is a habit that the classroom layout reinforces. Though rows are common for a reason—there are only so many ways to arrange 30 desks in the available floor space—they can make it difficult for students to hear or see one another well. I wonder if teacher-student mini-dialogues are inevitable given these physical constraints. One solution I've tried is removing the desks from the classroom entirely and rearranging the chairs into small discussion circles of six to eight students facing one another. I've found that students are more willing to speak freely in this kind of setting.

Even in small groups, students must be shown how to explore ideas with their classmates. One skill they must learn is how to listen to other students' conclusions and respectfully explore why they might not be true. For example, in the lesson on rocks and lava, some students snorted derisively when one of their classmates mentioned thinking that the melted thymol was water. Though these students did a good job marshaling their

evidence and reasoning for why thymol was not liquid water, their eagerness to show that they were right prevented them from exploring the legitimate reasons why someone might think melted thymol was water.

To address this kind of problem, I frequently have students lead whole-class discussions on topics that we've explored thoroughly. I step to the back of the room as a student speaker steps to the front, usually to share a model or an explanation using a document camera. I post a few reminders for the speaker: "Orient your paper so it's visible on the screen," "Face the class," "Point to your model or explanation as you talk." These prompts function as a safety net preventing the speaker from freezing up in front of his or her peers. The final speaker prompt is always to open the topic for group discussion. This gives the student speaker the power to invite and moderate discussion among classmates.

The listeners also have prompts to help them respond in respectful ways. I give the student audience the following sentences to help frame their responses:

- I agree with _____ because _____.

- I heard you say _____. Is that right?

- I disagree with _____ because _____.

- Could you say that again?

- I heard you say _____, but I was wondering _____.

Over time, students gradually develop productive habits of discourse to use in their small-group discussions when an adult moderator is not present. Although nurturing meaningful discourse is a long process, and many of my efforts have felt unproductive, I know our hard work is paying off when I hear a student say effortlessly and entirely unprompted, "I don't think what you're saying makes complete sense. Could you tell me what you meant by _____?"

Looking Forward by Looking Back: Teachers as Practitioners of Science

I began my teaching career a decade before the publication of the *National Science Education Standards* (NRC 1996) and the *Benchmarks for Science Literacy* (AAAS 1993). I still remember my delight in reading these documents. The authors captured essential ideas and ways of knowing in the sciences. These 20-year-old ideas are still key to the vision developed in the *Framework*. For example, on page 9 of the *Benchmarks*, the authors write the following:

"Scientific inquiry is more complex than popular conceptions would have it. … It is far more flexible than the rigid sequence of steps commonly depicted in textbooks as 'the scientific method.' More imagination and inventiveness are involved in scientific inquiry than many people realize, yet sooner or later strict logic and empirical evidence must have their day. … If students themselves participate in scientific investigations that progressively approximate good science, then the picture they come away with will likely be reasonably accurate. But that will require recasting typical school laboratory work."

Having "students themselves participate in scientific investigations that progressively approximate good science" remains part of the vision of the *Framework* and the *NGSS*. Why didn't this idea take root earlier? The report *Science Teachers' Learning* (NRC 2015) suggests a possible explanation: that "science teachers need rich understandings of [practices, core ideas, and crosscutting concepts]. Perhaps equally important, **they need to be able to engage in the practices of science themselves** and know how to situate this new knowledge in learning settings with a range of students" (p. 100, author emphasis).

My understanding of how to engage in the practices of science and with the core ideas of science and crosscutting concepts came from my experience doing science with a mentoring scientist. It didn't come from *talking about* how to do science, but from actually *doing* it. Having experienced doing science as a scientist has been critical to making my classroom function productively. Managing a classroom, keeping up with the inchoate blizzard of ideas kids have, and capitalizing on students' contributions requires tremendous daily effort. My experience as a novice scientist all those years ago and as a

mentoring scientist in my own classroom since then gives me the confidence, skill, and enthusiasm to keep going.

I wonder if the time has come for us to expect science teachers to be scientists, or at least practitioners of science who teach with the mindset that they are investigating with their students and mentoring them as they engage in the practices of science. We expect music teachers to be musicians and art teachers to be artists. Why not expect science teachers to be scientists?

References

American Association for the Advancement of Science (AAAS). 1993. *Benchmarks for Science Literacy*. New York: Oxford University Press.

Colson, M., and R. Colson. 2016. Planning for NGSS-based instruction: Where do you start? *Science Scope*. 83 (2) 16–18.

Colson, R., and M. Colson. 2016. *Learning to read the Earth and sky: Explorations supporting the* NGSS. Arlington, VA: NSTA Press.

National Research Council (NRC). 1996. *National Science Education Standards*. Washington, DC: National Academies Press.

National Research Council (NRC). 2012. *A framework for K–12 science education: Practices, crosscutting concepts, and core ideas*. Washington, DC: National Academies Press.

National Research Council (NRC). 2015. *Science teachers learning: Enhancing opportunities, creating supportive contexts*. Washington, DC: National Academies Press.

NGSS Lead States. 2013. *Next Generation Science Standards: For states, by states*. Washington, DC: National Academies Press. *www.nextgenscience.org/next-generation-science-standards*.

Reiser, B. 2014. Designing coherent storylines aligned with the NGSS for the K–12 classroom. Paper presented at the national conference of the National Science Education Leadership Association, Boston. *www.academia.edu/6884962/Designing_Coherent_Storylines_Aligned_with_NGSS_for_the_K-12_Classroom*.

Mary Colson *is an eighth-grade Earth science teacher at Horizon Middle School in Moorhead, Minnesota. During her 24 years of classroom teaching, she has engaged students with the big ideas of Earth science through authentic explorations in the lab and outdoors. She is coauthor of the 2016 NSTA Press book* Learning to Read the Earth and Sky: Explorations Supporting the *NGSS. Additionally, Colson was a member of the* NGSS *writing team, has served on NSTA's Council of District Directors, and is a former president of the Minnesota Science Teachers Association. In 2008, she received the Medtronic Foundation Science Teaching Award for Middle Level Science in Minnesota. Colson has a BS in geology from Allegheny College and an MS in geology from the University of Tennessee-Knoxville. She completed her teacher certification as a Lyndhurst Fellow at the University of Tennessee-Knoxville. The author can be contacted by e-mail at* mcolson@moorheadschools.org.

CHAPTER 4

Constructing Explanatory Arguments Based on Evidence Gathered While Investigating Natural Phenomena in a Secondary Biology Classroom

Tricia Shelton

In this chapter, we examine the use of explanatory arguments for empowering students to construct their own understanding of scientific concepts. In the Shelton Class, high school students are guided by their own questions as they gather and make sense of data, share their ideas, attempt to persuade others of those ideas, and allow their conceptual models to be critiqued, debated, and revised. In this way, students build understanding by using science and engineering practices to figure out explanations for the phenomena that they observe. This chapter presents a framework students can use to construct arguments that they then revise over time within a community of science learners that extends beyond the classroom.

Introduction

What is your vision for your students? What do you hope they can attain as a result of their education? As an educator, what is your reason for doing what you do? These important questions amplify the need for and the beauty of *A Framework for K–12 Science Education* (*Framework*; NRC 2012) and the *Next Generation Science Standards* (*NGSS*; NGSS Lead States 2013). The *Framework* describes three principal goals of science education: to cultivate scientific habits of mind, to engage in scientific inquiry, and to reason in a science context. By realizing these goals, students will be empowered to act as critical consumers, problem solvers,

and communicators—roles that are necessary for "their individual lives as citizens in this technology-rich and scientifically complex world" (NRC 2012, p. 10).

The science classroom is ideal for cultivating habits of mind that are essential for preparing students to function responsibly in society, because students must use those habits as they engage in the practice of scientific argumentation. The *Framework* and the *NGSS* provide us with a set of research-based objectives that lay out the pathways through which students best learn science, and now it is up to educators to implement the science and engineering practices that support student learning. The Shelton Class methodology has shown

Table 4.1. Principal Goals of Science That Connect to Practice of Argumentation

Principal Goals of Science Education		
Scientific Habits of Mind Valuing evidence in the creation of a body of knowledge as well as in communication; providing and receiving evidence-based critique	**Engaging in Scientific Inquiry** Question generation serving as guideposts for learning; student-generated questions motivating study and persistence	**Reasoning in a Scientific Context** Communicating evidence with a connection and justification of science ideas and concepts; evaluation of competing arguments to determine their merits

that construction of explanatory arguments, combined with the social interaction that accompanies this practice, empowers students to take an active role in the process of thinking and learning (see Table 4.1).

The Shelton Class is actually a combination of classes taught at Boone County High School from 2014 to 2016 and at Randall Cooper High School from 2016 to 2017. Both schools are part of the Boone County Schools district in northern Kentucky. At the time of the Shelton Class, the district served over 20,000 students, of whom more than 1,200 were English language learners who spoke more than 50 different native languages. Forty percent of Boone County students received either free or reduced lunch.

Navigating Biology Through the Practice of Argumentation

If we consider the *NGSS* to be our destination, then we can think of the instructional activities we use to arrive there as points on a family trip. Within the Shelton Class methodology, it follows that explanatory arguments are our global positioning systems, allowing us to explore stops on our journey without losing our way. Explanatory arguments help students become proficient in scientific practice while exploring core science concepts in a safe and supportive environment. Crosscutting concepts further help to organize and connect students' thinking about key ideas, ensuring that three-dimensional learning drives students' learning path.

By sharing the Shelton Class framework here, we hope to provide an example of a path that leads to the destination envisioned by the *NGSS*. It is our hope that students who have journeyed with us arrive at this destination able to act as critical consumers, problem solvers, and communicators beyond the classroom as they participate in social, cultural, and economic affairs throughout their adult lives.

Scientific Explanations and Arguments

Science is a way of explaining the world, and scientific literacy is an underlying goal of both the *Framework* and the *NGSS*. Using explanatory arguments and engaging in the process of argumentation are central practices that unite all subject areas. Although the *NGSS* separates the two skills, in reality they depend on each another. To construct explanations of their world, students must also be able to use evidence to formulate and support logical arguments. Through scientific argumentation, students work to understand observable phenomena, communicate their understanding of core ideas and concepts, and convince others of their conceptualizations (Berland and Reiser 2011). When students engage in argumentation, they develop a framework for intertwining knowledge and practice much as the science community does.

In an *NGSS* classroom, students are continually working to explain phenomena or develop solutions to problems. Because students all have varying types and levels of prior knowledge, it is not surprising when they do not immediately agree on a single possible explanation or proposed solution. However, students who develop proficiency in the skill of argumentation are able to consider multiple hypotheses by evaluating the strength of the empirical evidence offered. Focusing on the practice of argumentation helps students to develop an *evolving scientific body of knowledge* as well as to provide evidence-based explanations for processes occurring in the natural world. "The argumentation and analysis that relate evidence and theory are essential features of science; scientists need to be able to examine, review, and evaluate their own knowledge and ideas and critique those of others" (NRC 2012, p. 27). When students use argumentation as the path to explanation, they are acting and thinking like scientists.

Conceptualization and Communication Through the Practice of Explanation

A key focus in both the *Framework* and the *NGSS* is using science to make sense of the world or to design solutions to problems. More specifically, the focus is on active processes, such as figuring out explanations and solutions, as opposed to passive processes, such as hearing or reading about ready-made explanations. In addition to *using the science to figure out phenomena*, another necessary shift in instructional practices involves the use of inquiry to guide the process of discovery and learning. Inquiry is the pursuit of science knowledge by developing explanations of phenomena, and it has been at the center of reform efforts for the past two decades. In the *NGSS*, inquiry remains a focus.

Questions still guide the journey, but now students are engaged in figuring out the answers for themselves by gathering and making sense of data, sharing their ideas, and persuading others. These practices allow students to develop an understanding that can be critiqued, debated, and revised within their peer groups. Our desired endpoint is an explanation of phenomena that has been built, revisited, and revised over time within our own scientific community of learners.

Figure 4.1 summarizes our unit on photosynthesis and conversion of energy to mass. Notice that inquiry is driving the overall learning about the anchor phenomenon. Questions also drive each of the four lessons indicated by the rows in the graphic organizer. In order to investigate each question, students engage in the science and engineering practices and use the crosscutting concepts indicated in the second column. For example, to understand the core science idea of photosynthesis, students may engage in the practices of planning and carrying out investigations and analyzing and interpreting data while using the crosscutting concepts of Patterns and Cause and Effect to organize their thinking about the data they collect using a leaf

Figure 4.1. High School Biology Science Unit Storyline: Seed to Tree

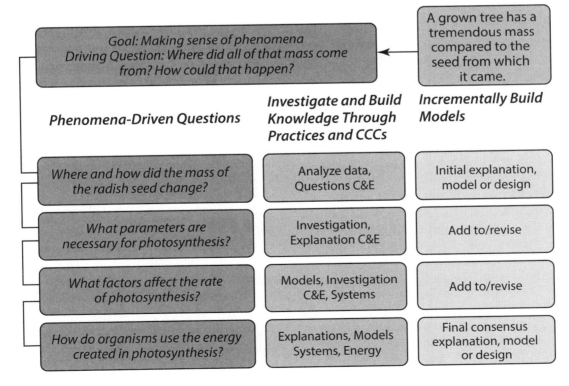

disc assay. Once students generate and think about this empirical evidence, they can communicate their new evidence-based understanding of photosynthesis and the concept of matter and energy flow using the practice of explanation. As students progress through the unit, they are working to explain the unit phenomena (how a seed becomes a tree) and building an understanding that evolves with each lesson.

As shown in Figure 4.1, discovery in our classroom is centered on student questions about phenomena. The desire to provide an explanation for these questions and communicate that explanation to others is the reason students are collecting data and thinking about claims and evidence. As students work to make sense of their data using the crosscutting concepts to organize and connect their thinking, they are continually communicating their understanding to others and persuading others of that understanding through the practice of argumentation. For student-generated questions to guide the lessons illustrated in Figure 4.1, the Shelton Class made use of a visual organizer that we referred to as a Driving Question Board.

Figure 4.2. Radish Seeds

1.5 g of seeds were placed on moist paper towels under the following conditions for one week. The final biomass (dried in the oven overnight so no water is left) measured in grams as follows. Explain the results.

Ebert-May (2003)

At the beginning of the unit, students were presented with the data in Figure 4.2, and they began their investigation by noting what they did and did not understand about the information. Students were then asked to consider both the anchor phenomenon and the radish-seed data as they constructed questions that needed to be answered using science concepts. Students wrote their questions on notecards, which were then displayed on a poster board. These questions about the phenomena drove our study. As the unit progressed, students used notecards of different colors to add questions to the board (see Figure 4.3). The cards on

Figure 4.3. Driving Question Board With Notes

the left side of the board are yellow to show our initial questions, and the cards on the right side are red to show questions that were added as the students engaged in the investigative process.

Instead of only learning the discrete facts of photosynthesis, students learned about the topic within the context of an observable phenomenon. By using the storyline shown in Figure 4.1, students in the Shelton Class were eventually able to figure out for themselves the process through which a seed becomes a grown tree. This ability to construct explanations by analyzing evidence from class experiences using the crosscutting concepts as thinking tools is the motive behind three-dimensional learning. Communicating lesson-level understanding and describing the progress they are making toward this goal gives students opportunities to engage in argumentation and allows teachers to monitor students' progress and guide them toward three-dimensional understanding.

Phenomena-driven instruction makes learning relevant, because the questions being investigated are based on the interest and curiosity of the students being taught. However, in order for these lessons to be instructionally productive, the ideas that students attempt to explain must be directly mapped to the disciplinary core ideas of the *NGSS*. To align our investigative topics with the core ideas, I grouped the cards from the driving questions board into subsets of related inquiries. To answer these inquiries, the students had to connect the evidence generated through classroom investigations to specific core ideas. After making these connections, students were ready to use argumentation to communicate, persuade, and critique. During the seed-to-tree investigation, for example, student questions were grouped under these larger inquiries:

- Why and how did the mass of the radish seed change? (*NGSS*: PS1.B, LS1.C)

- What parameters are necessary for photosynthesis? (*NGSS*: PS3.D, LS1.C)

- What factors affect the rate of photosynthesis? (*NGSS*: PS3.D, LS1.C)

- How do organisms use the energy created in photosynthesis? (*NGSS*: PS3.D, LS2.B)

Each of these topics was then coupled with a lesson-level investigation of observable characteristics associated with the phenomenon. Once they have generated evidence, students use crosscutting concepts to analyze how it connects to the core ideas. Using this method, the teacher can set the stage for explanatory arguments that promote a deep understanding of core ideas and concepts. Despite discovering answers during each lesson, unanswered questions always remain, and the need to further their investigations to address those questions drives students to continually increase their depth of understanding. This cycle of generating and answering questions continues until students have created an explanation or a model that demonstrates mastery of the overall concept.

Classroom Culture and Argumentation

Questions serve to frame our investigative journey. As indicated in Figure 4.1, students use a combination of the eight science and engineering practices to make sense of the phenomenon being investigated. During this process, students are engaged in argumentation, persuasion, and critique as they think about the data and communicate evidence-based explanations of lesson-level phenomena. However, in order for students to engage in this type of learning, teachers must first create an environment that supports argumentation.

As a high school teacher, I am sensitive to the challenges presented when students critique one another's responses. For students to engage in the type of discourse that supports science learning, norms must be established for valuing and debating arguments without judgment. In the Shelton Class, we spend time at the start of our year establishing a community of respect and defining expectations for day-to-day interactions.

We define our community as an environment in which students feel valued as individuals as well as connected to their teacher and to one another. Students believe that they have something to offer the community and that the community has something to offer them. When students feel valued and connected, they can develop the confidence that allows them to acquire competence. As a class, we begin our journey by establishing an environment that supports inquiry, and I make certain to clearly explain to my students how acting and thinking like scientists within our classroom community will prepare them for their futures regardless of the path they choose.

After establishing the real-world application of the skills we are developing, we move to the process we will be using to develop those skills. Here we engage in two main modes of work: The Alone Zone provides opportunity for independent reflection, and Science Talk provides an opportunity to improve understanding through discourse. To support these processes, we developed the norms shown in Figure 4.4 and Figure 4.5 (p. 34). As a result, visitors to our classroom observe evidence-based communication and argumentation, and this ability, more than understanding any one particular phenomenon, is the true product of our endeavors.

Figure 4.4. The Alone Zone

Alone Zone Norms Internal Dialogue

I will honor internal dialogue time by
- remaining silent,
- responding honestly, and
- focusing my thinking (consider and commit).

In the Alone Zone, students are instructed to think first (consider) and then commit (actually write a response or draw a model) to the prompt or problem provided. This commitment on paper more deeply engages students in the reflective and cognitive process. Alone Zone time usually comes prior to Science Talk and oral argument, giving students time to think through their own ideas and evidence before sharing them. Shelton Class students report that this "in my own head" prep time is essential for facing the risks involved in sharing ideas and reasoning with a group.

Figure 4.5. Science Talk

Collective Conversation Norms: Science Talk

- We communicate evidence-based thinking.
- We honor each other and all our voices.
- We are honest but honorable to differences in opininon.
- We have the right and responsibility to contribute to the conversation, making our thinking public to move the learning of others.
- We commit to the group by contributing to the learning of others.

The purpose of Science Talk, which is summarized in Figure 4.5, is to improve the group's conceptualization of the topic being discussed by sharing the thoughts and questions of all members of the group. Because students want to feel valued by the group, once they understand that this process benefits both themselves and others, they will enthusiastically engage in the discussion so long as they receive the proper supports. Crosscutting concepts ensure that both individual thinking and collective thinking organize and connect to the phenomenon under study and the science ideas that can be used to explain it.

Once the class norms for the Alone Zone and Science Talk have been established, students form "Mini Think Tank" groups that support the practice of argumentation. These processes work together to facilitate discourse and support learning for all students. While engaging in discourse, students must first clarify their thoughts about phenomena or data, then share them and support them with evidence that stands up to critique from the group. Collectively, the Mini Think Tank considers multiple pieces or lines of evidence from competing arguments in an effort to explain a phenomenon. For these interactions to succeed, students must be taught to describe their understanding and support it with evidence, to encourage participation from all group members, and to respectfully give and receive critiques. Students in the Shelton Class are not focused on who is sharing the "right" or "wrong" answer but rather on who is sharing evidence-based thinking to advance the discussion and group understanding of the scientific processes being considered.

Multiple Vehicles of Argumentation

Once a safe and respectful classroom environment has been established, students are ready to engage in three-dimensional learning by communicating arguments and explanations in a variety of formats. In the Shelton Class, students engage in argument from evidence through speaking or Science Talk, through writing, and through digital media. Because all three of these vehicles are rooted in the same scientific habits of mind, they work together to support students in acting and thinking like scientists and in developing a deep understanding of science ideas and concepts. Despite the similarities among these activities, students see them as offering very different opportunities. The nuances of each keeps students motivated and engaged, while the common foundation in scientific thinking supports an increased depth of understanding. In a typical week, students engage in argument from evidence and communication of understanding through all three vehicles. Together, these communicative activities help students become proficient with the practices of argument and explanation.

Individually, each type of communicative activity provides its own unique benefits to student growth and achievement. For example, when students produce oral arguments by engaging in Science Talk, they are sharing their evidence-based thinking, explaining their understanding of phenomena, and asking questions of one another. Within their Mini Think Tanks, students assume specific roles that keep sessions focused and productive. The leader is responsible for keeping the group on topic and for encouraging the group to move on when necessary and try reaching a conclusion about the explanation of a phenomenon. The facilitator supports each student in sharing his or her ideas or understanding by questioning and probing for evidence. Students often deliver multiple tentative explanations of phenomena or pieces of evidence to form a common explanation for debate. The summarizer listens closely to the various explanations and redirects the discussion when necessary by summarizing what's most relevant. Students receive scaffolding in all three roles, and they rotate roles throughout the course.

These supported Science Talk sessions allow students to become comfortable with topic-specific vocabulary and learn the social language of the scientific community. They also allow students to distill the information that they have gathered during their investigations into useful chunks and to clarify their thinking through communication. All students benefit from the models and conceptualizations shared by their peers as the groups engage in oral argument from evidence to advance their shared understanding.

In addition to oral argumentation, students also regularly communicate arguments through writing. While oral argumentation builds on the collaborative nature of the science and engineering practices, written argumentation gives students the opportunity to synthesize information from Mini Think Tank sessions and evaluate the data gathered from class investigations to communicate their own understanding through argument. Additionally, writing allows students to document their evolving understanding as they refine their models to account for new information. In the Shelton Class, we record our initial thoughts about the phenomenon at the beginning of the unit, then revisit and expand on the explanation as we continue our class investigations. The oral and digital-media arguments allow students to help one another make sense of the evidence and connect related science concepts, but, at the end of each unit, the students all individually prepare a report or article that describe their unique interpretations of the knowledge and understanding that they have gained.

Science Talk and Mini Think Tanks both play a valuable role in training students to engage in scientific dialogue, but these discussions are limited to the classroom. Although many students bring unique ideas and experiences to these discussions, the students in one particular school can be expected to have similar stores (and gaps) in prior knowledge. These commonalities limit their ability to effectively engage in the type of critique that allows for conceptual growth. If we are to achieve 21st century learning goals, then it is appropriate to make use of 21st century tools. It is for this reason that the Shelton Class incorporated a third vehicle of argument, the use technology, into our collaborative process. By utilizing video and other media to engage in explanatory argument with a larger audience, the students can further refine not only their communication skills but also their ability to engage in critical thinking

or critique and to distill information into component concepts.

To gain this wider audience, students create two- to three-minute video arguments or explanations, identifying the data that most strongly support their claims and discarding any weak or unnecessary evidence. The goal is to persuade others through scientific reasoning. In our seed-to-tree storyline, we followed our discourse about the radish seed data with a leaf disc assay to answer the lesson-level question, "What are the parameters for photosynthesis?" After engaging in the science and engineering practices of planning and carrying out an investigation and analyzing and interpreting data, students entered the Science Talk phase of discourse to make sense of their evidence and develop further understanding.

To construct explanations, students must engage in argument from evidence (Berland and Reiser 2011). In the seed-to-tree investigation, the video argument is an extension of this construction of explanation, because students attempt to persuade others of their newly developed understanding of photosynthesis by explaining their interpretation of the leaf disc assay evidence. Video arguments allow students to give and receive evidence-based critique through peer review so that they can truly act and think like scientists as they distill and evaluate evidence.

Science Is Social: Peer-to-Peer Critique and Review

Once students have established a community of scientific collaboration within the four walls of the classroom, they are ready to extend that community beyond the classroom in a class-to-class peer review framework that further supports the practices of argumentation and explanation. With the help of BenchFly, our business and community partner, the Shelton Class has created a network that connects classrooms around the country. This peer review framework provides real and relevant contexts for discourse, interaction, and evidence-based critique. Students use video and other media to collaborate and communicate with peers through two-minute video arguments and explanations that can be recorded on computers, tablets, or cell phones. During this phase, the students in both classrooms are the constructors of the argument or explanation.

Students then send the videos to their connected classroom peers for evidence-based review. After sharing their products, students in both classrooms become peer reviewers responsible for providing evidence-based critique to their partners in the other classroom. Students provide feedback by evaluating claims, analyzing the rationale and evidence presented in the videos, and challenging conclusions. They engage in Science Talk and then use peer feedback to refine and strengthen their argument or explanation. My colleague Amanda Meyer has noted that, by expanding the audience, the video peer-review process required students to consider a wider range of scientific implications, leading them to develop models that could be applied in broader contexts. The process also helps to ensure that students are recognized by others for developing expertise.

The peer-review process allows students to mirror the practices of the scientific community. In doing so, students learn to act and think like scientists. They learn to determine which types of evidence are accepted by the scientific community, as well as how to communicate, critique, and validate ideas. These interactions within their own community of scientific collaboration encourage students to develop the scientific literacy and scientific identity that will continue to define their understanding of the natural world long after they complete the course.

Conclusion and Lessons Learned

Empowered, informed, independent: These are some of the words I use to describe the characteristics that I wish to cultivate in my students. When they leave us after 12 years of education, I want them to be well prepared to take the next step in any education or career path they choose. All students should have an equal opportunity to be successful in their adult lives. Educators who understand and use the practice of argumentation as an intersection of knowledge building, evaluation, and communication help students to develop both scientific literacy and employability skills. Classrooms that use the *NGSS* to guide instruction and assessment while focusing on equity are grounded in research about the most effective ways to support student learning. The practice of argumentation gives students many opportunities to develop language, literacy, and 21st-century skills.

In the Shelton Class, students work to develop an explanation of phenomena or to solve a problem. They use the data and evidence we generate or utilize

in class to explain phenomena observed in the real world. Engaging in argument from evidence through discourse, writing, and digital media enables students to act and think like scientists as they develop literacy and to leverage opportunities beyond the classroom.

Translating the *NGSS* into instruction in my classroom has felt like the rebirth of an already rewarding career as a science educator. I share some of my experiences from the Shelton Class here in hopes of allowing others to see the incredible potential and engagement of an *NGSS* classroom. That said, shifting practice is undoubtedly a challenge. Here are three key lessons my colleagues and I have learned as we've made these shifts in our classrooms.

Lesson 1: Change is hard—on everyone. Knowing and communicating the reasons we believe as we do is essential. There are many differences between traditional and *NGSS*-aligned science education. Methods of grading, types of student work, and the pace and structure of the classroom are all affected. These changes can be accepted much more readily by all involved, including students, parents, administrators, and educators, if an explicit plan to communicate new expectations and benefits of making changes is in place. Among these benefits are the opportunities for post-secondary success granted by the development of critical-thinking skills and scientific literacy.

Lesson 2: Have patience while learning to act as a facilitator rather than an authority. In the *NGSS* classroom, the teacher is no longer the "sage on the stage" delivering knowledge but rather the "guide on the side" offering students space to build understanding, confront misconceptions, and experience the empowerment of figuring things out independently. Instead of expecting the students to be patient and listen, it is *we* who must be patient and listen to *them* as they talk through their misconceptions, address oversights, and reformulate ideas. Although students are expected to take an active role in this work, they need guidance and support, and it is our role to provide that mentorship by designing experiences that cause students to confront their own misconceptions. Educators must provide artifacts and resources that lead students to their own, valid conclusions. Students may carry misconceptions for a while, and it can be hard to refrain from interfering. Although students are used to accepting authoritarian corrections to their knowledge base, retention is fleeting if they don't have the opportunity to construct a personal understanding of material. By contrast, when students decide for themselves that their conceptual or mental model of a phenomenon no longer makes sense and decide to alter their own understanding, then that understanding will persist. After three years in an *NGSS* classroom, I can say with confidence that students will confront and correct misconceptions on their own if they are provided with opportunities to learn through conversation and debate with others.

Lesson 3: Partner with students. The students in the Shelton Class are full partners. It was their initial idea to leverage digital media in multiple ways to enhance our *NGSS* experience. Accepting their contribution required me to embrace the use of technology with which the students had more expertise than I did. This culture of learning from and with one another has been a cornerstone of our success.

Learning in an *NGSS* classroom and designing *NGSS*-aligned experiences require many changes that must be continually implemented over time. Educators must accept that things do not always go as planned, and they must have faith that failure will ultimately lead to growth. It is this growth mindset that has been the key to the success of our program—and more important, the success of the students.

References

Berland, L. K., and B. J. Reiser. 2011. Classroom communities' adaptations of the practice of scientific argumentation. *Science Education* 95 (2): 191–216.

Ebert-May D., J. Batzli, and H. Lim. 2003. Disciplinary research strategies for assessment of learning. *Bioscience* 53 (12): 1221–1228.

National Research Council (NRC). 2012. *A framework for K–12 science education: Practices, crosscutting concepts, and core ideas.* Washington, DC: National Academies Press.

NGSS Lead States. 2013. *Next Generation Science Standards: For states by states.* Washington, DC: National Academies Press. *www.nextgenscience.org/next-generation-science-standards.*

Tricia Shelton *is a high school biology teacher at Randall Cooper High School in Union, Kentucky. She has 22 years of experience teaching and serving as a teacher leader. She has a BS in biology and an MA in teaching. In 2014, Shelton received an NSTA Distinguished Teaching Award. Driven by her passion to help students develop critical and creative thinking skills, she also strives to help other teachers embed these skills in their science instruction by serving on the NSTA NGSS Advisory Board, working with educators as an EQuIP rubric trainer, and acting as an NSTA Professional Learning Facilitator. She can be contacted by e-mail at* tdishelton@gmail.com.

CHAPTER 5

Assessment That Reflects the Three Dimensions of the *NGSS*

Susan German

In this chapter, we address the importance of explicitly teaching crosscutting concepts, adding scientific argumentation to assessment, and redesigning assessments to move students toward attaining the performance expectations outlined in the *Next Generation Science Standards* (*NGSS*; NGSS Lead States 2013). Crosscutting concepts improve learning by helping students recognize connections between seemingly unrelated ideas. We also describe suggestions for implementing scientific argumentation and writing performance assessments.

Purpose and Overview

As a science teacher in a small, rural school, I have been working with eighth-grade students for the past 20 years. Despite the small size and remote location of my school, I was lucky enough to receive 18 months of high-quality professional development from Mid-Continent Research for Education and Learning (McREL) on the topic of assessment. By the end of this training, I was able to prepare worthwhile prompts and use student responses to guide my instruction. However, I did not have a source for high-quality assessment questions. My search for a better resource soon led me to Page Keeley's *Uncovering Student Ideas* series. Over the years, I attended several of Keeley's training sessions on the use of formative assessment probes or questions. These training sessions, plus many related readings, improved my ability to assess student knowledge. Around this same time, the publication of *A Framework for K–12 Science Education* (*Framework*; NRC 2012) started a change in the methods used to

teach and assess science students, regardless of school setting.

The *Framework* established the need for students to directly experience both science and engineering practices in the classroom: "Students cannot appreciate the nature of scientific knowledge without directly experiencing and reflecting on the practices that scientists use to investigate and build models and theories about the world. Nor can they appreciate the nature of engineering unless they engage in the practices that engineers use to design and build systems. The experiences provided to students should be as authentic as possible in order to prevent students from getting by with only rote memorization of facts or procedures. By creating an immersive experience, students will come to view science as an … iterative process of empirical investigation, evaluation of findings, and the development of solutions. Similarly, students will come to view engineering as a process of developing and improving a solution to a design problem" (Pellegrino, Wilson, Koenig, and Beatty 2014, p. 30).

To accurately assess the type of learning, it is important to establish three priorities. First, assessment should be used as a tool for learning. Skillful use of formative assessment to inform classroom instruction is a proven method of improving student conceptual attainment. Second, the teacher must meticulously determine which ideas students have actually learned, and to what extent. The *Framework* encourages educators to focus on depth of understanding, because regurgitation of facts does not equate to critical thinking and applied learning. Third, the assessment must support all types of learners. To meet these three goals, I had to make many changes to my classroom assessment practices. I had to become more explicit in teaching crosscutting processes, add scientific argumentation as a practice to be assessed, and redesign my assessments to move students toward meeting *NGSS* performance expectations.

The Importance of Crosscutting Concepts

Crosscutting concepts are those that can be applied to disciplines outside of science. Such concepts include patterns; scale, proportion, and quantity; cause and effect; systems and systems models; energy and matter; structure and function; and stability and change. Studying these helps students make connections to other subjects (NRC 2012, p. 84).

After reflecting on crosscutting concepts as described in the *Framework*, I initially felt that they were nothing new. However, upon further reflection, I realized that I was not straightforward about the presence of these concepts during lessons. Over the past couple of years, I have found that the more explicit I am when teaching crosscutting concepts, the better my students perform. Prior to doing so, I was failing to harness the power of these concepts to help students develop effective problem-solving skills.

The importance of explicitly teaching crosscutting concepts was made clear to me by the results of a formative probe that I used to assess the standard MS-ESS1-1, "Develop and use a model of the Earth-Sun-Moon system to describe the cyclic patterns of lunar phases, eclipses of the Sun and Moon, and seasons" (NGSS Lead States 2013). I had used "Shorter Days in Winter" from Keeley and Sneider's formative assessment book *Uncovering Student Ideas in Astronomy* (2012):

Mrs. Moro's students checked the newspapers every morning for the times of sunrise and sunset. They used this information to determine the number of hours of daylight. The class started this project in September, and by November they could see that the days were getting shorter and shorter. The students asked their families and neighbors to explain why days get shorter as winter approaches in the North. Here are the ideas they came to class with the next day:

Frank: "My mom says it's because of daylight saving time."

Jubal: "My sister said Earth's tilt causes the Sun to be farther away in the winter."

Sybil: "My father thinks the angle of sunlight must be the cause."

Carter: "My brother says the Sun moves across the sky faster in winter."

Wendy: "My neighbor thinks the Sun's path in the sky gets shorter in winter."

Which student came to class with the best idea? _____. Explain why you think that is the best idea.

The lessons were centered on finding patterns in photoperiods, lunar phases, and seasons. Before using this probe, I thought I had covered everything, and my eighth-grade students could easily recite the pattern of the photoperiod in each major band of the Earth's latitudes. However, after checking the student

Table 5.1. Student Responses to the Assessment Prompt

Answer Options for "Shorter Days in Winter"	Percentage of Eighth-Grade Students Selecting Each Option
Frank	11%
Jubal	38%
Sybil	36%
Carter	4%
Wendy	11%

Table 5.2. Cause and Effect

Cause		Effect
Frank:	"My mom says it's because of daylight saving time."	Daylight savings time shifts the hours of daylight, but does not increase/decrease the hours of daylight.
Jubal:	"My sister said Earth's tilt causes the Sun to be farther away in winter."	The tilt changes the angle the Sun's rays strike the Earth's surface.
Sybil:	"My father thinks the angle of sunlight must be the cause."	The angle of the Sun impacts how Earth's atmosphere and surface is heated.
Carter:	"My brother says the Sun moves across the sky faster in winter."	The rate the Sun moves across the sky remains the same during all seasons.
Wendy:	"My neighbor thinks the Sun's path in the sky gets shorter in winter."	The Sun reaches a lower altitude in the northern hemisphere during the winter than it does during the summer.

responses, their scores proved an unpleasant surprise (see Table 5.1).

With the results in hand, it was time to analyze them and determine why students weren't meeting my expectations. The first possibility I considered was that my lessons may not have been sufficiently aligned to my assessment. My lessons were three-dimensional in nature. Students used the science and engineering practice of making models to manipulate photoperiod data and construct graphs, and they used the disciplinary core idea of seasons along with the photoperiod data to apply the crosscutting concept of determining patterns. It seemed clear that the lesson did match the assessment.

Next, I reflected on the wording of responses to determine whether they may have been confusing to students. I noticed that Jubal's response included the phrasing "Earth's tilt causes it to be farther away" and that Sybil's used the phrase "angle of sunlight." I considered the possibility that students had simply selected responses that contained phrases they recognized from the lessons rather than thinking through each choice. After reviewing the portion of the _Framework_ that discusses crosscutting concepts, I confirmed that the performance expectation from the _NGSS_ stated that students should use patterns. However, as I reviewed the list of concepts, it occurred to me that I had focused on identifying and working with patterns to the exclusion of all other concepts. My students could identify patterns, but they were unable to predict how any of

the causes mentioned in the response choices may have resulted in them.

The next day, we began working on a cause-and-effect chart that listed each of the response choices in the "cause" column (see Table 5.2). Students copied the chart into their notes, and we discussed the effect of each response, but our discussion did not go so far as to eliminate incorrect answers or reach a consensus. Instead, I left it to the students to evaluate the chart on their own and determine the best answer.

A day later, without any forewarning, I handed students a new copy of the formative assessment probe. The results showed a major change in student thinking: 92% of the students chose the best answer (see Table 5.3). By explicitly teaching the crosscutting

Table 5.3. Student Responses on Retest

Answer Options for "Shorter Days in Winter"	Percentage of Eighth-Grade Students Selecting Each Option
Frank	0%
Jubal	6%
Sybil	2%
Carter	0%
Wendy	92%

concept of Cause and Effect, I had achieved an 81% increase in student performance in a single day.

Incorporating Crosscutting Concepts Into Formative Assessment

When designing instruction, educators need to incorporate as many opportunities for students to interact with all concepts, including crosscutting concepts, evaluated in a summative assessment. Though there are several available options for integrating crosscutting concepts into daily instruction and formative assessment, the three I use the most are Crosscutter Cards; Two-Tier Question Probes; and the science, technology, engineering, and math (STEM) teaching tool Practice Brief 41: Prompts for Integrating Concepts into Assessment and Instruction (Penuel and Van Horne 2016).

Crosscutter Cards. This strategy, from Page Keeley (2015), is a formative classroom assessment technique, or FACT, in which students learn to connect seemingly unrelated science concepts. The cards I prepare for students define crosscutting concepts and provide examples of closely associated science and engineering practices. First, I ask students to choose a crosscutting concept to describe what they've just studied. Next, I ask them to think of another, related concept that they have recently learned. For example, when studying power, students may choose Scale, Proportion, and Quantity as a crosscutting concept, which they may then relate to the concept of density.

Two-Tier Question Probes. This strategy is a slight modification of one of Keeley's. Students first answer a prompt, usually in the form of a multiple-choice question, then provide the reason or "rule" they used to select their response using a crosscutting concept. If the crosscutting concepts are new to students, choose the one they should use in their answer for them.

Practice Brief 41: Prompts for Integrating Crosscutting Concepts into Assessment and Instruction. This tool lists several sentences or prompt starters for educators to use organized by crosscutting concept. For example, after graphing data, students are often asked to write a conclusion. With the prompts, I can scaffold the process by asking students questions related to patterns, such as "What does the pattern of data you see allow you to conclude from the experiment?" and "What do you predict will happen to [variable] in the future? Use the pattern you see in the data to justify your

answer" (Penuel and Van Horne 2016). Each suggestion represents a small change to current materials that raise the importance of crosscutting concepts to students by deepening their learning.

Adding Scientific Argumentation to Assessment

Prior to the release of the *Framework* and the *NGSS*, constructed-response questions were typically explanatory in nature (e.g., "How do earthquakes occur?" "How do the hours of daylight change during the course of a year?"). As a teacher, I learned to specify the types of responses that I wanted my students to provide. For example, to ensure that student responses demonstrated extended thought, I would ask, "What are three ways an aquarium can be used as a model for a pond?" instead of "How can an aquarium be used as a model of a pond?"

I used to think that specificity equated to good assessment, but I now realize that my prompts neither engaged my students in learning nor required complexity of thought. The shift to classroom application of science and engineering practices has created a need to expand assessment methods to evaluate higher-level abilities, such as use of scientific argumentation, if students are to meet the objectives of the *Framework* and the *NGSS*.

The practices of engaging in argument from evidence and of constructing explanations (for science) and designing solutions (for engineering) can seem very similar. Engaging in argument requires students to evaluate claims, resolve questions, and investigate ideas, and constructing explanations requires students to explain phenomena by applying accepted scientific theories or other related information. When creating an assessment, I sometimes find it difficult to distinguish between the two. To help keep them straight, I use the following explanation with my students: "Arguments answer 'How do you know?' and explanations answer 'Why is it so?'"

However, even with the addition of scientific argumentation, assessments still often focus on only two of the three dimensions of science learning—usually a science and engineering practice plus a disciplinary core idea. Assessing the third dimension requires precision when eliciting responses. With enough forethought, it is possible to develop a performance task that effectively assesses a disciplinary core idea, a crosscutting concept, and student ability to engage in argument from evidence.

My first attempts at adding scientific argumentation to a summative assessment were disastrous—I had to go back and rework almost all aspects of the unit. I made changes to my lesson presentation by incorporating explicit instruction and scaffolds for writing a claim, supporting it with evidence, and providing a rebuttal. I pared the science content I presented down to essential components described in a simple scenario that students could investigate, allowing them to focus on the skill of argumentation. Once students improved performance in that area, I introduced more challenging core science concepts.

Eventually, after several years, I also made changes to my grading criteria by adding use of crosscutting concepts to the scoring guide. However, rather than define the crosscutting concept that I expect students to use, I prefer to let the students choose one on their own and determine how it relates to the ideas they are describing. By adding this requirement to the scoring guide, I emphasize the importance of crosscutting concepts in science, and by writing out their thought process, students practice metacognition.

The following prompt and the scoring guide in Table 5.4 provide an example of an assessment that evaluates all three dimensions of teaching and learning:

Table 5.4. Scoring Guide

Component	Score		
	4	3	2
Accuracy of claim	Makes a scientifically correct claim and completely catches the essence of the investigation	Makes a scientifically correct claim and partially catches the essence of the investigation	Makes a scientifically incorrect claim
Sufficiency of evidence	Provides more than two pieces of evidence and makes a rebuttal	Provides two pieces of evidence	Provides one piece of evidence
Quality of evidence	Makes an appropriate and adequate explanation completely based on interpretation of investigation data	Makes an appropriate and adequate explanation partially based on interpretation of investigation data	Makes an inappropriate and inadequate explanation or reports data as evidence
The relationship between claim and evidence	Makes a strong and sophisticated connection between claim and evidence; Uses at least one croscutting concept	Makes a strong and sophisticated connection between claim and evidence	Makes a weak connection between claim and evidence
Presents a rebuttal or another claim	Rebuttal is coherent and provides appropriate evidence and explanation for disagreement	Rebuttal is somewhat incomplete with explanation only partially based on evidence	Makes a weak rebuttal with incomplete explanation not based on evidence
Multimode representations	Uses more than one mode (text) in explaining the concept(s) and it is tied to the text	Uses more than one mode (text) in explaining the concept(s) but does so separately from text	Uses only one mode (text) to explain the concept in writing
Audience language	Language is appropriate, easy to understand, and meets the demands of the audience	Although clearly aware of audience, the writer only ocasionally speaks directly to that audience	Does not consider the audience's language
Score			

A group of three students were discussing their favorite skate park that has a half-pipe. The half-pipe has a starting ramp 10 meters in height. The opposite half is also 10 meters in height. Assuming no friction, the skater that starts at the 10-meter height loses zero energy as they travel back and forth. The three students discuss what would happen if the skate park was located on the Moon and Jupiter. Below are their claims:

Samuel said, "Skaters on the Moon will travel farther than 10 meters on the second ramp because there is less gravity."

Thomas said, "The amount of starting energy the skater increases with gravitational pull which will cause the skater to travel faster."

Susan said, "The mass of the skater or the starting height does not impact the speed of the skater. The amount of gravity of the Moon and Jupiter impact the speed of the skater."

Choose the student you agree with the most and use their claim and construct an argument to support your choice. Remember, evidence includes trends of data and observations that support your claim. The simulation, Energy Skate Park, allows you to change the setting of the skate park to include Earth, Moon, and Jupiter. The simulation will also allow you to quantify the total energy, potential, and kinetic energy in addition to changing the mass of the skater.

When you compare the task above with examples of the constructed-response questions that I used prior to shifting my instruction, there are several noticeable differences. Factors such as the complexity of the task, the need for students to undertake a multistep process, and the integration of more than one concept all add depth and allow for a variety of responses. Another key factor is the requirement that students build their own explanation for the hypothesis that they choose to support.

This type of task can be applied to a wide variety of scenarios, many of which may be proposed by the students themselves. For example, students often wonder how various activities might be different if participants were on the Moon rather than on Earth. In cases such as this, where it would be difficult to measure or gather data, computer simulations can help students investigate the relevant core concepts. Energy Skate Park, created by the University of Colorado Boulder (2017), allows students to learn about the conservation of energy by allowing them to manipulate a virtual skater on tracks, ramps, and jumps. As the skater moves, the simulation provides data about changes in kinetic, potential, thermal, and total energy. As a model, the simulation demonstrates the effects that the laws of physics would have on the skater on Earth, on the Moon, and on Jupiter. By manipulating the simulation, students can observe effects that they could never witness firsthand and can then use these observations to figure out answers to their questions. The simulation also lets students choose from a variety of approaches as they design their experiments, thus engaging in the practice of planning and carrying out investigations as well as that of developing and using models. Further adding to the complexity of students' tasks, the simulation allows them to fully investigate claims that prove incorrect. Students who choose to investigate an incorrect claim will find they cannot collect data to support it, allowing them to develop their knowledge through meaningful work.

In the Energy Skate Park example, Energy and Matter is the most apparent crosscutting concept. Students consider the total energy of the system and the ways in which potential and kinetic energy change as the skater moves. By changing the location of the skate park, students learn about the influence of gravity on total energy. They are also able to increase and decrease the mass of the skater to determine the effects that the change has on total energy. Because the skate park is a system, students can also learn how defining the system affects the definition of the total energy.

As this example illustrates, three-dimensional assessment requires a lot of thought and effort from both students and teachers, but in the end results in a deep understanding of complicated concepts.

Creating Assessment Tasks

When I first took it upon myself to revise my assessments for three-dimensional learning, I spent a lot of time reading and rereading standards, performance expectations, and evidence statements. It took a lot of effort and a lot of revisions, but eventually I developed the following process for designing performance tasks.

I begin by identifying a task. I glean ideas from activities, news stories, and conversations with students, other educators, or members of the community. For example, the skate park lesson developed after students asked "What if?" questions during class. Their questions were complex, and at first I was tempted to redirect them to simpler ones that would be easier to investigate and explain. As educators, we are trained to present material in the most concise manner possible and to simplify information if necessary. However, when we oversimplify science concepts or streamline scientific investigations, students do not get opportunities to think critically about the material and achieve depth of understanding. When students investigate complex situations, such as in the Energy Skate Park simulation, they are required to use multiple science concepts, such as energy and gravity, and consider their interactions while synthesizing information to formulate an answer. Once I've determined a task of sufficient complexity, I add instructional scaffolds as needed to help students retrieve and synthesize data. These scaffolds often consist of a line of questions that move students toward a target.

By working through these considerations, I can create a clearly structured task that engages students in three-dimensional learning by having them combine content knowledge pertaining to several concepts and acquired from multiple sources and then apply that knowledge to a new situation.

The next step is to align the task to the standards. I accomplish this by identifying the underlying science concepts, science and engineering practices, and cross-cutting concepts demonstrated during the performance of the task. If any concepts or practices listed as a part of our unit goals are absent from the task, I attempt to revise the prompt, add additional parameters, or request further explanations or analysis. The Energy Skate Park example is aligned to the following two standards: MS-PS2-4, "Construct and present arguments using evidence to support the claim that gravitational interactions are attractive and depend on the masses of interacting objects," and MS-PS3-5, "Students who demonstrate understanding can construct, use, and present arguments to support the claim that when the kinetic energy of an object changes, energy is transferred to or from the object." Because the standards are quite broad, it is often impractical to completely capture them in a single performance task. Although three-dimensional assessment bundles multiple standards into a single evaluation, it is unlikely that each standard will be assessed fully within the bundle. I do not consider this to be a shortcoming of the standards or of the assessment task. Instead, I prefer to view assessment as a gauge that shows where students are in their progress toward eventually mastering the standard.

The last step in the task-creation process is to reflect and refine. Long ago, I decided that no assessment item is ever perfect. After administering an assessment, I use student feedback and my own observations to improve the structure of the question or task. The K–8 *NGSS* evidence statements produced by Achieve, Inc. can be helpful at this stage, providing additional detail about the information that students should know and the skills they should be able to demonstrate. The evidence statements for standard MS-PS3-5 note the observable features of student performance desired by the end of the course:

1. *Supported claims:* Students make a claim about a given explanation or model for a phenomenon. In their claim, students include ideas that when the kinetic energy of an object changes, energy is transferred to or from that object.

2. *Identifying scientific evidence:* Students identify and describe the given evidence that supports the claim, including the following when appropriate: (i) the change in observable features (e.g., motion, temperature, sound) of an object before and after the interaction that changes the kinetic energy of the object, and (ii) the change in observable features of other objects or the surroundings in the defined system.

3. *Evaluating and critiquing the evidence:* Students evaluate the evidence and identify its strengths and weaknesses, including: (i) types of sources; (ii) sufficiency, including validity and reliability, of the evidence to make and defend the claim; and (iii) any alternative interpretations of the evidence and why the evidence supports the given claim as opposed to any other claims.

4. *Reasoning and synthesis:* (a) Students use reasoning to connect the necessary and sufficient evidence and construct the argument. Students describe a chain of reasoning that includes the following conclusions: (i) Based on changes in the observable features of the object (e.g., motion, temperature),

the kinetic energy of the object changed. (ii) When the kinetic energy of the object increases or decreases, the energy (e.g., kinetic, thermal, potential) of other objects or the surroundings within the system increases or decreases, indicating that energy was transferred to or from the object. (b) Students present oral or written arguments to support or refute the given explanation or model for the phenomenon. (NGSS Lead States 2013)

Here are the observable features required by MS-PS2-4:

1. *Supported claims:* Students make a claim to be supported about a given phenomenon. In their claim, students include the following idea: Gravitational interactions are attractive and depend on the masses of interacting objects.

2. *Identifying scientific evidence:* Students identify and describe the given evidence that supports the claim, including (i) the masses of objects in the relevant system(s) and (ii) the relative magnitude and direction of the forces between objects in the relevant system(s).

3. *Evaluating and critiquing the evidence:* Students evaluate the evidence and identify its strengths and weaknesses, including (i) types of sources; (ii) sufficiency, including validity and reliability, of the evidence to make and defend the claim; and (iii) any alternative interpretations of the evidence, as well as why the evidence supports the given claim as opposed to any other claims.

4. *Reasoning and synthesis:* (a) Students use reasoning to connect the appropriate evidence about the forces on objects and construct the argument that gravitational forces are attractive and mass dependent. Students describe the following chain of reasoning: (i) Systems of objects can be modeled as a set of masses interacting via gravitational forces. (ii) In systems of objects, larger masses experience and exert proportionally larger gravitational forces. (iii) In every case for which evidence exists, gravitational force is attractive. (b) To support the claim, students present their oral or written argument concerning the direction of gravitational forces and the role of the mass of the interacting objects. (NGSS Lead States 2013)

According to the above statements, students need to be able to construct a claim that describes the ways in which energy transferred to or from the object affects its kinetic energy, and they must also demonstrate an understanding of the concept that gravitational forces are attractive and dependent on the masses of the interacting objects. In the Energy Skate Park example, these proficiencies are demonstrated to varying degrees of completeness. Additionally, students need to identify and use scientific evidence, both to support their own claims and to refute claims made by other students. Lastly, students must demonstrate their ability to use reasoning to connect the evidence. Connected ideas from the Energy Skate Park example include the interactions between systems of objects and gravitational forces, the ways in which increases or decreases in the kinetic energy of an object affect potential energy, and the constant nature of the total energy of the system. I use the evidence statements to determine the criteria of my scoring guide. Once I am satisfied that the lesson and scoring guide require students to demonstrate the desired proficiencies, all that remains is to format, review, and distribute the materials.

Final Thoughts

During my teaching career, I have seen much new research about factors that affect student learning. As with all new knowledge, flexibility and a willingness to change is needed to make use of these advances. However, one truth will always remain, in both the world of education and the world of science: The perfect solution does not exist. A lesson or assessment that worked flawlessly with one class might need to be adapted for the next class.

Assessment is a critical component of teaching and learning not only because analyzing the data informs lesson planning and execution but also because watching students proudly demonstrate their newly acquired skills and knowledge motivates me to continue my efforts to provide the best instruction and assessment possible. This is a goal that can never be fully accomplished, because classroom assessment is an ever-evolving body of knowledge. Though three-dimensional learning has created the potential for students to attain deeper levels of understanding, it has also created a need for assessments that align with the improved outcomes. Assessing in three dimensions has sparked my creativity and improved my instruction, and it is my sincere hope that science teachers across the country will find the process equally as rewarding.

References

Keeley, P. 2015. *Science formative assessment: 50 more strategies for linking assessment, instruction, and learning.* Vol. 2. Arlington, VA: NSTA Press.

Keeley, P., and C. I. Sneider. 2012. *Uncovering student ideas in astronomy: 45 new formative assessment probes.* Arlington, VA: NSTA Press.

National Research Council (NRC). 2012. *A framework for K–12 science education: Practices, crosscutting concepts, and core ideas.* Washington, DC: National Academies Press.

NGSS Lead States. 2013. *Next Generation Science Standards: For states, by states.* Washington, DC: National Academies Press. *www.nextgenscience.org/next-generation-science-standards.*

Pellegrino, J. W., M. R. Wilson, J. A. Koenig, and A. S. Beatty. 2014. *Developing assessments for the Next Generation Science Standards.* Washington DC: National Academies Press.

Penuel, W., and K. Van Horne. 2016. Practice Brief 41: Prompts for integrating crosscutting concepts into assessment and instruction. STEM Teaching Tools. *www.stemteachingtools.org/brief/41.*

University of Colorado Boulder. 2017. PhET interactive simulations. *https://phet.colorado.edu/en/simulations/category/new.*

Susan German *has 24 years of experience teaching math and science in Hallsville, Missouri. Although most of her experience has been teaching grade 8, she has worked with students from grades 6–12. She holds National Board certification and has earned a BS and MS in science education, a BA in chemistry, and a certification as an educational specialist in educational technology with an emphasis on learning systems design and development. Additionally, she has been honored with NSTA's Distinguished Teaching Award and the Science Teachers of Missouri Outstanding Service Award. She has held leadership positions on the Board of Directors for both of these organizations. German can be contacted by e-mail at susangermanscienceteacher@gmail.com.*

2

Professional Development Strategies That Support the Implementation of the *Framework* and the *NGSS*

CHAPTER 6

Promising Professional Learning: Tools and Practices

Rodger W. Bybee, James B. Short, and Dora E. Kastel

In this chapter, we address the importance of instructional materials and professional learning by clarifying what is new and different about contemporary state standards, emphasizing the critical link between instructional materials and professional learning, introducing the characteristics of transformative professional learning, and describing several tools and practices that can be used in professional learning to identify, select, or design instructional materials aligned with the *Next Generation Science Standards* (*NGSS*; NGSS Lead States 2013).

When states, school districts, and schools establish new standards for science education, it is essential to provide teachers with the knowledge, skills, instructional materials, and tools they will need to implement those standards. The influence of standards rests with teachers and the teaching practices that they employ in their classrooms. Ultimately, changes designed to improve science education must be implemented by teachers.

What Is New and Different About the *NGSS*?

The architecture of the *NGSS* differs significantly from that of prior standards for science education. In the *NGSS*, science and engineering practices, disciplinary core ideas, and crosscutting concepts form the three dimensions of learning. The objectives of this learning are clearly identified by means of performance expectations—statements of competency that describe the content and skills to be assessed following instruction.

A comprehensive instructional program should provide opportunities for students to develop their understanding of disciplinary core ideas through their engagement in science and engineering practices and their application of crosscutting concepts. This three-dimensional learning leads to eventual mastery of performance expectations. A high-quality science program should clearly describe or show how the cumulative learning experience works coherently to build scientific literacy.

The following innovations in the *NGSS* are hallmarks of current thinking about how students learn science, and they set a vision for future science education. These innovations will not only cause a shift in instructional programs and practices in American classrooms but also affect and refocus the efforts of curriculum developers and the design of comprehensive school science programs:

- Teaching to three dimensions—science and engineering practices, crosscutting concepts, and disciplinary core ideas

- Having students engage with natural phenomena and design problems to solve

- Introducing science practices and crosscutting concepts in ways that include engineering ideas and explore the nature of science

- Including units or yearlong programs based on coherent learning progressions

- Drawing connections to mathematics and literacy standards

In the report *Science Teachers' Learning: Enhancing Opportunities, Creating Supportive Contexts* (NRC 2015), Wilson et al. state that few K–12 science teachers have the expertise needed to teach the science required in the *NGSS*. On a more positive note, they also declare that the "*NGSS* represents a fundamental change in the way science is taught and, if implemented well, will ensure that all students gain mastery over core concepts of science that are foundational to improving their scientific capacity" (p. 1). The authors conclude that teachers will need new scientific knowledge and skills, new instructional approaches, and new instructional materials to implement *NGSS*-aligned strategies in their classrooms. "To enable teachers to acquire this kind of learning," they write, "will in turn require profound changes to current systems for supporting teachers' learning across their careers, including induction and professional development" (p. 2). Here are the report's recommendations:

- Take stock of the current status of learning opportunities for science teachers.

- Design a portfolio of coherent learning experiences for science teachers that attend to teachers' individual and context-specific needs in partnership with professional networks, institutions, and the broader scientific community as appropriate.

- Consider both specialized professional learning programs outside of school and opportunities for science teachers' learning embedded in the workday.

- Design and select learning opportunities for science teachers that are informed by the best available research.

- Develop internal capacity in science while seeking external partners with science expertise.

- Create, evaluate, and revise policies and practices that encourage teachers to engage in professional learning related to science.

- Explore the potential of new formats and media to support science teachers' learning.

Instructional Materials Matter

Choice of instructional materials can have as much of an effect on student learning as improvements to teacher learning (Chingos and Whitehurst 2012). Unfortunately, the current marketplace offers limited examples of high-quality, well-aligned instructional materials. Curriculum reform must therefore be accomplished through a systemic approach that requires educators to develop new instructional materials and learn new ways of using them. Putting curriculum reforms into practice is a difficult and demanding process that requires a vision of reform, support for change, collaboration among teachers to learn, and leadership at different levels (Anderson 1995). Reform-oriented instructional materials incorporate contemporary research on learning and challenge teachers to think differently about learning and teaching content knowledge. In addition to outlining what to teach, these materials must provide support for instructional activities and research-backed teaching methods.

Transformative Professional Learning

Relevant, ongoing professional learning can help teachers to use innovative instructional materials effectively. Transformative professional learning challenges teachers' beliefs about and understanding of content, teaching practices, and contemporary views of learning. Supporting teachers with a mix of reform-oriented, standards-based instructional materials and effective professional learning focused on transforming their beliefs and practices contributes to substantial improvement in instruction. Districts need to work with outside partners to develop better support systems that include such resources as professional learning facilitation, instructional coaching, and designs for adult learning. To access these resources, districts should consider partnering with nonprofit organizations with expertise in professional learning, school improvement, instructional design, and disciplinary content.

Professional learning experiences designed to support the implementation of new instructional materials should employ the same instructional methods as are to be used with students (Loucks-Horsley et al. 2010). These experiences must also challenge teachers' current beliefs about learning and teaching science. Thompson and Zeuli (1999) offer the following five key features of transformative professional learning:

- Create a level of cognitive dissonance that is high enough to disturb, in some fundamental way, the equilibrium between teachers' existing beliefs and practices and their experience with subject matter, students' learning, and teaching.

- Provide time, contexts, and support for teachers to think in order to enable them to work at resolving the dissonance through discussion, reading, writing, and other activities that promote the crystallization, externalization, criticism, and revision of their thinking.

- Ensure that the dissonance-creating and dissonance-resolving activities are connected to the teachers' own students and context.

- Provide a way for teachers to develop a repertoire for practice that is consistent with the new understandings that teachers are building.

- Provide continuing help in the cycle of (1) identifying the new issues and problems that will inevitably arise during initial implementation, (2) deriving new understandings from them, (3) translating these new understandings into performance, and (4) recycling ideas and reworking activities to align them with the new vision or adapting successfully employed activities to apply to additional content. (pp. 355–357)

Leading Professional Learning

It is not enough that these leaders of professional learning be good teachers themselves; they must also be prepared to work with adult learners and able to coordinate professional learning at the school and district

levels. It is critical that districts develop expertise in building the capacity of leaders of professional learning. In many cases, they will need to develop internal capacity while also seeking external partners with both content expertise and knowledge about understanding the change process, research on learning, and effective design of adult learning experiences that mirror those we want to see them provide to students.

Strategies learned during professional learning must be able to be applied on a large scale. The current prevailing method of scaling up teacher development experiences does not support the deep understanding that they must achieve before successfully implementing instructional innovations in the classroom. Too many professional learning opportunities focus on strategies that may work well on a small scale when led by expert facilitators but that cannot be easily replicated if educators lack the requisite knowledge, skills, and expertise to do so with fidelity. In the worst cases, we send teachers to professional learning and then expect them to turn around and teach their colleagues what they learned just as effectively. This is ridiculous. Nothing in education is "turnkey." Important aspects of the learning experience will always be lost when the person leading the session is only passingly familiar with how to use the instructional materials.

Transformative professional learning must engage educators in thinking about how students learn, how coherent instructional materials support learning, how specific teaching strategies support the research on learning, and how we assess student learning to determine deep understanding of content and ability to problem solve.

Tools That Support Professional Learning and *NGSS*-Aligned Instructional Materials

Schools and districts need help selecting the highest quality *NGSS*-aligned instructional materials, and a variety of tools are available for the purpose. However, before selecting any materials, teachers must know how to assess their effectiveness. The Educators Evaluating the Quality of Instructional Products (EQuIP) rubric for science lessons and units, developed by Achieve, Inc. in partnership with NSTA, is designed to help educators do just that and is available for download at *www.nextgenscience.org/resources/*

equip-rubric-lessons-units-science. EQuIP provides criteria for evaluating the degree to which lessons and units are aligned to *NGSS* as well as a process for using the rubric to review existing materials and provide criterion-based feedback and suggestions for improvement. Additionally, Achieve, Inc. has developed a Peer Review Panel trained in using the EQuIP rubric and finding examples of high-quality *NGSS*-aligned resources.

The EQuIP rubric was designed to evaluate a lesson or a unit that includes instructional tasks and assessments aligned to the *NGSS*. The rubric works best when multiple reviewers examine the same materials, allowing for discourse and deeper professional learning. For discourse to be constructive, it is essential that reviewers begin the process with at least an initial understanding of the *NGSS*.

Reviewers begin by familiarizing themselves with the student and teacher materials along with the intended *NGSS*-aligned disciplinary core ideas, cross-cutting concepts, science and engineering practices, and performance expectations for the lesson or unit. The vision of science education articulated in the seven conceptual shifts of Appendix A of the *NGSS* is captured within the three *NGSS*-aligned categories of the rubric, which are devoted to three-dimensional learning, instructional supports, and monitoring student progress. Once familiar with the materials and the intended *NGSS* alignments, reviewers engage in a deliberate process, going through the questions for each criterion within a category. EQuIP supports the vision of the *Framework* in that an instructional sequence is rooted in an explanatory question aimed at making sense of a phenomena or designing a solution to a problem.

Category I of the rubric begins the review process with questions focused on each of the three dimensions and learning centered at their nexus. When analyzing a unit, reviewers look for coherence across lessons and for connections to the *Common Core State Standards* for both mathematics and English language arts. Participants engaging in this work as a group will often start individually before coming together to discuss their evidence and reasoning, arriving at a consensus on evidence of quality (i.e., inadequate, adequate, or extensive), and discussing suggestions for improvement.

After Category I, reviewers may pause to consider whether their review is worth continuing. Lessons or units lacking sufficient evidence to show that they are aligned to the *NGSS* are labeled "not ready to review."

Categories II and III follow a similar process. In Category II, reviewers consider scaffolded questions about relevance and accuracy, building on students' prior ideas and interest, and differentiation over time. In Category III, they consider questions about formative assessments embedded in the lesson or unit and unbiased tasks or items they might employ to assess the suitability of the lesson or unit for furthering three-dimensional learning.

If not labeled as "not ready to review," materials can receive one of the three following overall ratings upon completion of the review process:

- Exemplifies the criteria in all three categories of learning
- Adequate overall, some improvement in one or two categories needed
- Significant improvement in one or more categories needed

The EQuIP rubric does not require consensus among team members but does emphasize discussion as a key component of the review process. By asking teachers to look for evidence of alignment at the element level in all three dimensions, the process deepens teachers' understanding of the *Framework* and the *NGSS*.

Teaching Educators to Select the Right Materials

Once reviewers have identified high-quality *NGSS*-aligned instructional materials, they need to teach classroom teachers how to do the same rather than rely exclusively on reviewers' evaluations. In fact, teaching teachers how to analyze and select such materials needs to be considered an essential professional development strategy. One resource for ensuring this is the Analyzing Instructional Materials Process and Tools (AIM) model developed by the Biological Sciences Curriculum Study (BSCS) and its partners at the K–12 Alliance at WestEd. The model, which has also been used in preservice courses, has been employed by the BSCS for over 15 years to help educators choose the right materials.

Originally designed as a professional development strategy to support curriculum implementation, AIM takes an inquiry-based approach consistent with a constructivist view of learning, focusing on asking questions, gathering information, and making decisions about instructional materials based on evidence. The model encourages teachers to think about the importance of instructional materials in the learning process for both students and teachers.

Working as a team, teachers and administrators begin the AIM process by completing a graphic organizer showing the conceptual flow of content in a unit. Next, they use rubrics to analyze evidence from materials related to the specific content, including assessment options and supports available to teachers. Following this detailed screening process, teachers and administrators field-test selected materials using additional tools to gather evidence on student understanding and teacher implementation.

A study (Short 2006) focused on a group of teachers using AIM found that they recognized the need for instructional materials to explicitly support concept development and the understanding of big ideas in science and the importance of including a "storyline" of content enabling students to make connections between ideas. The study also showed that the AIM process strengthened teachers' belief that students need to actually do science if they are to understand science content and raised their awareness of the power of high-quality materials to support inquiry-based learning through the application of science and engineering practices. Additionally, the study found that AIM reinforced the value of a student-centered approach to learning. The BSCS is currently working with WestEd and Achieve, Inc. to update AIM so that it more fully aligns with the *NGSS*.

Another option for helping teachers select *NGSS*-aligned materials is the Primary Evaluation of Essential Criteria for *NGSS* Instructional Materials Design (PEEC). In 2017, Achieve, Inc. released the PEEC, which builds on the criteria in the EQuIP rubric and focuses on evaluating the degree to which instructional materials are aligned to the *NGSS*. Before diving deeply into the materials of an instructional unit, PEEC participants use a simple tool to select those worth reviewing and then apply the EQuIP rubric to a parallel sample of units or instructional sequences across the program being reviewed. Materials found to have at least a "High-Quality Example, If Improved" rating are further examined for the types of learning opportunities they provide and their coherence across the lessons and units.

Both the PEEC and the revised version of AIM draw heavily from the recently released *Guidelines for the Evaluation of Instructional Materials for Science* (BSCS 2017), which outlines the research base for the criteria used in evaluation tools and their associated processes.

Developing High-Quality *NGSS*-Aligned Units

Teachers in many schools and districts are already beginning to implement the *NGSS* themselves by developing their own lessons and units. For some, this means a blended process of adapting old lessons and planning new ones. Educators need a clear process to develop coherent and effective three-dimensional lessons aligned to the *NGSS*.

The Five Tools and Processes for Translating the *NGSS* were designed to help professional development leaders work with teachers on curriculum, instruction, and assessment. In collaboration with BSCS and the K–12 Alliance at WestEd, the Gottesman Center for Science Teaching and Learning at the American Museum of Natural History in New York City developed and field-tested the Five Tools for professional development leaders. These tools are a timely and appropriate response to the challenges of translating the *NGSS* into classroom instruction and assessment, helping teachers to establish a meaningful context for understanding the *Framework* and the *NGSS* and start implementing changes in their classrooms. The Five Tools comprise a professional development curriculum that includes facilitation guides, slides, handouts, and templates, and is available for download at *amnh.org/ngss-tools*.

The main purpose of the Five Tools is to facilitate the translation of science concepts, practices, and performance expectations into multiple instructional sequences that form an *NGSS*-aligned unit. Additionally, the professional development curriculum includes suggestions for preparing more in-depth plans for instructional sequences and assessment tasks and for collecting evidence of student learning focused on a bundle of performance expectations. These processes help teachers plan for conceptual coherence.

Tools 1 and 2 begin the planning process, and Tools 3 to 5 engage teachers in design and development. In Tool 1, participants unpack a standards page by mapping out a conceptual and coherent blueprint for an instructional unit. This experience helps allay many teachers' apprehensions about the difficulty of implementing the *NGSS*. As participants begin planning their unit, they engage in close readings of sections from the *Framework* and the *NGSS* and begin adapting instructional sequences that they have previously used to them. These sequences are targeted to specific grades and topics, and teachers are encouraged to examine the progression charts for elements of the three dimensions that they can incorporate into their revisions. This process helps teachers to better understand both the vertical alignment of their unit to related concepts taught in other grades and horizontal alignment to the nature of science, engineering, and other topics.

Many teachers do not know which types of assessments are most effective in an *NGSS*-aligned environment. Tool 2 enables teachers to deconstruct a bundle of performance expectations from their Tool 1 blueprint and begin to plan for classroom evaluations. Participants are challenged to determine what counts as evidence of student learning and how to develop their own evidence of learning specifications that demonstrate a deep level of understanding of the science and engineering practices, crosscutting concepts, and disciplinary core ideas expressed in the targeted bundle of performance expectations.

In Tool 3, teachers analyze two different teaching scenarios to learn about the 5E Model, which familiarizes them with effective methods for facilitating student-centered learning focused on sense-making through phenomenon-based investigations or problem solving strategies. Tool 3 helps teachers develop a conceptual flow of the science content and refine the storylines of each instructional sequence within their unit.

In Tool 4, participants use Tool 1 (the unit blueprint) and Tool 3 (the 5E Model and conceptual flow) to outline more detailed instructional sequences for their unit by describing the teacher's and students' respective roles throughout. Finally, participants consider the activities they want to use in class using analysis guides to assess which ones fit the storyline of the sequence or need revising. At this stage, teachers consider students' prior knowledge and experiences and add in activities or revise plans connecting the lessons to crosscutting concepts in mathematics and English language arts that support three-dimensional learning.

In Tool 5, participants use the evidence of learning specifications identified in Tool 2 and the unit's performance expectations to create three-dimensional performance tasks that serve as summative assessments for each of the instructional sequences.

The Five Tools provide teachers with templates along with structured processes for designing three-dimensional units, instructional sequences, and classroom assessments. Teachers who use the tools come away with a deeper understanding of the three dimensions of learning, *NGSS*-related performance expectations, and an effective system for developing instruction with embedded formative assessments. Through this process, teachers gain experience using a constructivist, student-centered instructional model to create *NGSS*-aligned instructional sequences and units.

Conclusion

In this chapter, we explored promising professional learning strategies and emphasized the importance of assessing instructional materials for both quality and *NGSS* alignment. We also made a case for supporting professional development leaders by providing them with the tools and processes necessary for transformative learning to happen. Finally, we described four different sets of tools that support professional learning for *NGSS* alignment, all of them centered on identifying, selecting, and designing high-quality instructional materials aligned to standards.

The science education community has a responsibility to provide classroom teachers with opportunities for professional learning that complement the *NGSS*. Only in doing so will the community successfully address the challenges educators confront in implementing the reforms envisioned by the *Framework* and associated standards.

References

Anderson, R. D. 1995. Curriculum reform: Dilemmas and promise. *Phi Delta Kappan* 77 (1): 33–36.

Biological Sciences Curriculum Study. (BSCS). 2017. *Guidelines for the evaluation of instructional materials for science. http://guidelinesummit.bscs.org.*

Chingos, M., and G. Whitehurst. 2012. *Choosing blindly: Instructional materials, teacher effectiveness, and the common core.* Washington, DC: Brown Center on Education Policy, Brookings Institution. *http://brookings.edu/research/reports/2012/04/10-curriculum-chingos-whitehurst.*

Loucks-Horsley, S., K. Stiles, S. Mundry, N. Love, and P. Hewson. 2010. *Designing professional development for teachers of science and mathematics.* 3rd ed. Thousand Oaks, CA: Corwin.

National Research Council (NRC). 2012. *A framework for K–12 science education: Practices, crosscutting concepts, and core ideas.* Washington, DC: National Academies Press.

National Research Council (NRC). 2015. *Science teachers' learning: Enhancing opportunities, creating supporting contexts.* Washington, DC: National Academies Press.

NGSS Lead States. 2013. *Next Generation Science Standards: For states, by states.* Washington, DC: National Academies Press. *www.nextgenscience.org/ next-generation-science-standards.*

Short, J. 2006. Selecting instructional materials as professional development to support teacher change. In *The cornerstone-to-capstone approach: Creating coherence in high school Science.* Colorado Springs, CO: BSCS.

Thompson, C., and J. Zeuli. 1999. The frame and the tapestry: Standards-based reform and professional development. In *Teaching as the learning profession*, ed. L. D. Hammond and G. Sykes. San Francisco, CA: Jossey-Bass.

Rodger W. Bybee *was executive director of the BSCS until he retired in 2007 and today works as an education consultant. Before his work at BSCS, he was executive director of the Center for Science, Mathematics, and Engineering Education at the NRC. Bybee has taught at the elementary through college levels and has contributed to science education through publications and various projects, including work on the 1996 and 2013 national standards. Bybee was honored to receive the Distinguished Service Award and the Robert H. Carleton Award from NSTA. He can be contacted via e-mail at* rodgerwbybee@gmail. com.

James B. Short *is the program director for Leadership and Teaching to Advance Learning at the Carnegie Corporation of New York. His work in philanthropy focuses on supporting teachers and school leaders and developing high-quality instructional tools for implementing new standards. Before this work, he was a director in the education department at the American Museum of Natural History in New York City and science curriculum coordinator for Denver Public Schools. His classroom experience includes teaching secondary science and graduate education courses.*

Dora E. Kastel *is a leader of professional development programs at the American Museum of Natural History, focusing primarily on helping science teachers implement* NGSS *and* Common Core State Standards *literacy strategies. Before her work at the museum, she was a middle school science and math teacher. Kastel is pursuing a PhD in science education at Teachers College, Columbia University. She is a 2016–2017 CADRE fellow.*

CHAPTER 7

Preparing Teachers to Successfully Implement the Three Dimensions of the *NGSS*

David T. Crowther and Susan Gomez Zwiep

This chapter briefly outlines the challenges faced by leaders seeking to design and execute professional development that enables science educators to successfully implement the *Next Generation Science Standards (NGSS; NGSS Lead States 2013)* in their classrooms. The chapter explores recent literature on *NGSS*-aligned professional development and describes the adoption process of two *NGSS* states.

Introduction

The Next Generation Science Standards *were first introduced to my fifth-grade team in 2013. Administration informed us that this was where we were going in our science curriculum and that there was no looking back. The only issue was that we had absolutely no idea how to make sense of the standards and were found scrambling to create materials that would work for the unclear curriculum we were now required to use.*

—Ashley, a teacher in a graduate-level advanced science methods course

The above quote is from a master teacher who is pursuing graduate-level work in science education and who has a track record of being a highly effective teacher. Yet Ashley's initial teacher preparation left her in a quandary because she was not trained to write lessons and implement instruction based on the more recent standards.

Like all science teachers in states that have adopted the *NGSS* or *A Framework for K–12 Science Education* (*Framework*; NRC 2012), Ashley needs high-quality professional development. The *NGSS* requires teachers to apply strategies that differ vastly from both the way they were taught as kids and the methods they learned in their teacher preparation courses. This consequence of educational reform is not unique to the sciences: The *Common Core State Standards for Mathematics* (*CCSS*; NGAC and CCSSO 2010) also require teachers to use methods of teaching and learning mathematics with their students that they themselves most likely have not experienced, and *CCSS* for English language arts requires teachers to emphasize nonfiction literature and literacy development within the context of other content areas much more than was traditionally the case.

The *NGSS* ask teachers to seamlessly combine what used to be three sets of standards into one—a skill set that is rarely developed in certification programs, perhaps in part because it works best when teachers have enough experience to see the connections between the standards and to know their students' abilities. Whereas the National Science Education Standards (NRC 2006) had three discrete sets of standards for content, process skills in science, and unifying themes in science, the *NGSS* synthesizes them all as a single set of performance expectations. This change requires teachers to think differently about science instruction and to move away from the content-only approaches of the past. According to Reiser (2013), instructional shifts in disciplinary core ideas need to "shift from teaching facts to explaining phenomena" (p. 4).

Reiser further explains that "inquiry is not a separate activity—all science learning should involve in engaging in practices to build and use knowledge" (p 7). This requires a shift away from the common practice of teaching scientific inquiry and the nature of science separately and apart from content, usually during the first lab of the semester. According to the *NGSS*, these concepts are best taught through the continuous use of science and engineering practices that help students discover content in the context of real-world phenomena. The *NGSS* use crosscutting concepts to seamlessly integrate disciplinary core ideas and science and engineering practices with connections to life beyond the classroom. This type of science learning is meaningful to students because they can relate it to the world around them.

Reiser summarizes the major instructional shifts presented in the *Framework* as follows:

1. "Lessons [are] structured so that student work is driven by questions arising from phenomena, rather than topics sequentially pursued according to the traditional breakdown of lessons.

2. "The goal of investigations is to guide construction of explanatory models rather than simply testing hypotheses.

3. "Answers to science investigations are more than whether and how two variables are related, but need to help construct an explanatory account.

4. "Students should see what they are working on as answering explanatory questions rather than learning the next assigned topic.

5. "A large part of the teachers' role is to support the knowledge building aspects of practices, not just the procedural skills in doing experiments.

6. "Extensive class focus needs to be devoted to argumentation and reaching consensus about ideas, rather than having textbooks and teachers present ideas to students.

7. "Teachers need to build a classroom culture that can support these practices, where students are motivated to figure out rather than learning what they are told, where they expect some responsibility for this work of figuring out rather than waiting for answers, and where they expect to work with and learn with their peers." (p. 11)

In *A Guide to Implementing the Next Generation Science Standards* (2015a), the National Research Council (NRC) further explores the instructional changes required by the *NGSS*. Table 7.1 summarizes many of these shifts and compares them to more traditional instructional techniques.

In a recent study by Horizon Research (Trygstad, Smith, Banilower, and Nelson 2013), less than 1% of K–2 teachers and less than 2% percent of teachers in grades 3–5 had taken any engineering courses. An astonishing 77% of K–2 and 69% of 3–5 teachers were not adequately prepared to teach engineering as part of their science instruction. By contrast, the study found that 26% of middle school science teachers and 61% of high school science teachers *have* had course work in engineering (Smith, Banilower, Nelson, and Smith 2013).

The NRC (NRC 2015) compared decades of research on professional development programs with the requirements of the *Framework* and the *NGSS* and found that the most effective programs shared the following characteristics:

1. Teachers' science content learning is intertwined with pedagogical activities such as analysis of practice.

2. Teachers are engaged in analysis of student learning and science teaching using artifacts of practice such as student work and lesson videos.

3. There is a focus on specific, targeted teaching strategies.

Table 7.1. Guide to Implementing the *NGSS*

Science Education Will Involve Less of	Science Education Will Involve More of
1. Learning disconnected from questions about phenomena	1. Systems-thinking activities and modeling to explain phenomena and to contextualize new ideas
2. Teachers providing information to the whole class	2. Students conducting investigations, solving problems, and engaging in discussions with teacher guidance
3. Teachers posing questions with only one right answer	3. Students discussing open-ended questions that focus on the strength of the evidence used to generate claims
4. Students reading textbooks and answering questions at the end of each chapter	4. Students reading multiple sources and developing summaries of information
5. Students completing worksheets	5. Students writing journals, reports, posters, and media presentations that offer explanations and arguments
6. Teachers oversimplifying activities for students who are perceived to be "less able" to do science and engineering	6. Teachers providing supports that allow all students to engage in sophisticated science and engineering practices

4. Teachers are given opportunities to reflect on and grapple with challenges to their current practice.

5. "Learning is scaffolded by a knowledgeable professional development leader.

6. "Analytical tools support collaborative, focused, and deep of science teaching, student learning, and science content." (NRC 2015, pp. 134–135)

The NRC goes on to summarize the following four characteristics of professional development in science education as especially effective:

1. Active participation of teachers as they analyze examples of effective instruction and student work

2. Focus on content

3. Alignment with district policies and practices

4. Length sufficient to allow repeated practice or reflection on classroom experiences

NGSS and California

As one of 26 states that helped to develop the *NGSS*, California formally adopted the standards less than six months after the final draft was released, on September 4, 2013. Many of California's scientists and engineers helped with the development and adoption of the *NGSS* and continue to support their implementation. The California State Board of Education (SBE) and Department of Education have also strongly supported the *NGSS*.

Previous California standards focused on recall of science facts rather than on how to apply these facts in the real world (California Department of Education 2000). Although an investigation and experimentation strand was included, it was isolated from the specific content standards. The *NGSS* place a greater emphasis on having students develop a deeper understanding of broad scientific concepts and introducing them to the workplace practices of scientists and engineers.

In California, we had to divide up the standards for grades 6–8 and assign them to specific grade levels. The Science Expert Panel that had recommended the state adopt the *NGSS* was charged with creating this sequence. The panel was composed of K–12 teachers, other educators, scientists, and business and industry representatives. After several days of analysis and debate, the panel decided on performance expectations to align with the *CCSS* at each grade level, allowing for the integration of science learning with English language arts and mathematics. For example, disciplinary core ideas (and associated performance expectations) that require at least an eighth-grade level of mathematical sophistication (e.g., those necessitating an understanding of time scale or natural selection) were assigned to grade 8. Core ideas and expectations were also assigned to each grade level to allow for articulation and integration across levels as well as to support the logical flow of content and increasing complexity of concepts in grades 5–8. In such a way, the life-science content developed each year from K–5 continues to develop unbroken in grades 7, 8, and 9, rather than each middle grade focusing on a different science discipline—as was the norm before adoption of the *NGSS*.

This model is consistent with those used by the ten highest-performing countries on international science tests. However, because the integrated curriculum represented a significant departure from previous practice in California, it was initially controversial, engendering significant rebuttal from the field. Ultimately, the SBE did acknowledge its preference for the fully integrated model. (Because California is a strong local-control state, the SBE continued to allow the use of the original discipline-focused model as an alternative.)

The *NGSS* Early Implementer Initiative

The *NGSS* Early Implementer Initiative helps districts develop the capacity to fully implement the standards in grades K–8 well ahead of the established timeline. The Initiative is funded through the S. D. Bechtel Jr. Foundation and developed by the K–12 Alliance at WestEd, in close collaboration with and with the support of the SBE, the California Department of Education, and Achieve, Inc. Recognizing the complexity of the *NGSS*, the initiative offers educators four years of related support. Eight districts and two charter management organizations (the latter paid for through the Hastings Quillin Fund) were selected to participate in the initiative. Selection was based on an existing commitment to K–8 science education, readiness to implement the *NGSS*, and fidelity to California's integrated sciences model for grades 6–8. A desire to represent the diversity of California schools in terms of demographics, location, and size also factored into the selection process. Collectively, the initiative serves approximately 200,000 students across California, offering professional development focused

on three components: a summer institute with on-site implementation and classroom-embedded lesson study cycles, grade-level lesson study sessions in professional learning communities (PLCs), and a leadership component.

Summer Institute

Five-day institutes are held each summer to provide content-rich, in-depth professional development. Teachers spend about half their time at the institute engaging in integrated-content experiences facilitated by a cadre made up of a university scientist, a K–8 teacher, and a science pedagogy expert (such as a university science educator or state or county science specialist). The other half of the institute focuses on pedagogy sessions for helping teachers make the conceptual shifts required by the *NGSS*. These sessions are cofacilitated by teacher leaders and a member of the K–12 Alliance professional development team. During the first year, the institute focused on developing leadership for the core leadership team of teachers and administrators. Then, during the second through fourth years, all participating teachers, teacher leaders, and administrators were invited to take part.

The intent of the cadre is both for teachers to experience an *NGSS* learning sequence and to provide a snapshot of the *NGSS* in the classroom. Rather than modeling activities that could be used in grades K–8, the content experiences are designed to challenge teachers at the adult level. Additionally, inquiry-based pedagogy is used to facilitate content learning as well as to have teachers experience being on the receiving end of them. Many teachers are not accustomed to intense learning environments, making it all the more important for them to provide content-rich learning experiences that include practice opportunities (Garet, Porter, Desimone, Birman, and Suk Yoon 2001; Jeanpierre, Oberhauser, and Freeman 2005). In the K–5 band, our cadre works in the areas of life science, physical science, and Earth and space science by focusing on one discipline each year and integrating the others as appropriate. Our 6–8 cadre combined these three science disciplines into an integrated storyline presented over the course of three years.

The content sessions present a three-dimensional view of science concepts. The core ideas are used to identify the learning goals, and each cadre then centers its week around a specific phenomenon by engaging teachers in relevant science and engineering practices and crosscutting concepts. Often, nature of science themes are also used to anchor the overall arc of the sessions. The intent of these learning experiences is to allow the cadre to develop a vision of three-dimensional *NGSS* instruction that they can use in the classroom. To accomplish this goal, participants build their understanding of the identified core ideas, science and engineering practices, and crosscutting concepts. By giving teachers the opportunity to develop a personal understanding of teaching strategies from the learner's perspective, we hope they will be empowered to pursue similar types of learning experiences in their own classrooms.

During the half of the summer institute devoted to facilitating the conceptual shifts required by the *NGSS*, teachers participate in sessions devoted to crosscutting concepts or science and engineering practices. They are also introduced to tools such as the 5E Instructional Model (Bybee et al. 2006) and Conceptual Flow (DiRanna et al. 2008) that can help them design and implement three-dimensional *NGSS* instructional sequences and learn how to use notebooks to help students make sense of their ideas.

Lesson Study Sessions

Summer gives teachers the time to dig deeply into new thinking about science, instruction, and the *NGSS*. However, the ideas that teachers bring away from the summer institute are often fragile, and teachers need time to reflect and process them before they're ready to introduce them in their own classrooms. For this reason, our professional development program continues during the academic year through grade-level specific, two-day lesson study cycles called teaching learning collaboratives (TLCs) (DiRanna, Topps, Cerwin, and Gomez-Zwiep 2009). During these cycles, three or four teachers and a trained facilitator from the K–12 Alliance work together to design a three-dimensional learning sequence using the 5E Instructional Model. The K–12 Alliance adds a step to this model asking teachers to identify expectations for students as well as a concept column that illustrates the development of a science concept from students' prior knowledge to the learning goal of the lesson sequence (Gomez Zwiep et al. 2011). The facilitator uses a protocol that pushes teachers to consider how core ideas, science

and engineering practices, and crosscutting concepts develop through the sequence of the lesson. All three of these components are identified in the concept column, and teachers use a separate column to predict student responses to each phase of the lesson. Due to the complexity of designing three-dimensional *NGSS* lessons, an entire day is devoted to planning.

On the second day, the K–12 Alliance facilitator debriefs the plan developed the previous day, and then the teachers work collaboratively to teach the lesson. The purpose of the debriefing is to provide time for reflection and revision if necessary. After the team teaches the revised lesson, a second debriefing takes place to evaluate the effectiveness of the revisions. Both debriefings use student work generated during the lesson as evidence.

Teams participate in two TLCs per year, with the level of sophistication and expertise demonstrated increasing as the teachers grow in their understanding of the *NGSS*. The ideas from the TLCs are also used at grade-level meetings and PLCs to solicit additional feedback from colleagues and to monitor and adjust instruction as needed.

Leadership Development

Our Early Implementer Initiative specifically requires the involvement of administrators and the development of teacher leaders. Each of our districts identified a subset of K–8 teachers to serve as leaders tasked with building on existing local talent to expand the capacity of each site during and following implementation of the initiative. Teacher leaders serve as advocates for the new standards in their respective schools, consulting with colleagues on reframing lessons, answering questions about the new standards, and working to ensure that the *NGSS* receive as much attention as other curriculum areas. Teachers who have been selected as leaders attend additional professional development sessions in January and June to support their development in this role. These sessions bring together teacher leaders from across the state and engage them in productive conversations about issues related to leadership, science teaching, and shifts related to the *NGSS*.

We required at least three administrators to commit to the initiative for their district to be included in the initiative. Administrators are crucial to the success of *NGSS* implementation, providing insight into the inner workings of each district or site and helping teachers effect change. Administrators can address their teachers' unique concerns, relieve anxiety about navigating the logistical complexities of their sites, and provide ongoing support outside of professional development sessions. Administrators were required to attend training sessions with their teacher teams as well as to attend separate sessions supporting their unique role in the initiative. Administrator-specific sessions explore such topics as conducting fair and effective observations of *NGSS*-aligned classrooms and selecting and using protocols to document evidence of *NGSS* implementation.

Lighthouse Districts for *NGSS* Implementation

Despite its complexity, the *NGSS* elegantly reflects an authentic view of science instruction. Our Early Implementation Initiative embraces the standards' complexity, providing teachers and administrators with in-depth, long-term, content-rich professional development to support their journey toward full implementation. Ongoing support for implementing lessons from the summer institute is offered through TLCs, grade-level meetings, and PLCs. By working in partnership with Achieve, Inc. as well as state-level organizations, lighthouse districts illuminate a path of reform that other local educational agencies may choose to follow. As implementation moves forward, our early implementers will continue to try out new tools, processes, assessments, and instructional materials, learning more and more about successfully implementing *NGSS*-aligned strategies.

NGSS in Nevada

In the summer of 2014, the Nevada State School Board (NSSB) voted to adopt the *NGSS* as the state standards for science. Like school boards in many states, the NSSB is required to modify National Standards for the state level, so it made a few minimal changes (removing the clarification statements from performance expectations; changing the color schema from orange, green, and blue to blue, silver, and lilac), called them the Nevada Academic Content Standards in Science (NVACSS), and submitted the standards to the legislature, which officially adopted them.

The Nevada legislature requires a traditional three-year implementation process for new standards. In the first year, teachers became familiar with the standards; in the second year, they use the standards to guide instruction; and in the third year, the state begins using a uniform assessment aligned to the new standards. This tight time frame poses a challenge, as does the fact that the state allocates minimal funds for implementation of the NVACCS. Additionally, districts in Nevada are largely in charge of designing and running their own professional development.

After the legislature passed the NVACCS, a team of science educators from across the state gathered to design a transition plan from the old state standards to the new ones. This team soon determined that the NVACCS would take a minimum of five years to implement and devised the following eight criteria for successful implementation:

1. Communication

2. Statewide capacity and networks

3. Professional learning

4. Instructional practices and shifts in classroom teaching

5. Instructional materials and curriculum

6. Assessment systems

7. Policy shifts

8. Data collection

Each of these criteria was given one of three labels: *awareness, implementation,* or *continuing.* Key stakeholders from across the state were engaged to help design the implementation plan, which can be found at *https://sites.google.com/rpdp.net/nvacssguide.*

In the first year of the adoption cycle, professional development was focused on learning about three-dimensional learning and content integration, reviewing the *NGSS* appendixes to familiarize teachers with core ideas, crosscutting concepts, science and engineering practices, and using phenomena to frame units of instruction and lessons. During this same period, the Regional Professional Development Program (RPDP) in southern Nevada combined online and face-to-face training to introduce over 300 teachers across the state to the most important aspects of the new standards (Vallett et al. 2016). The northwest branch of the

RPDP also worked with the University of Nevada at Reno to train educators in rural northern Nevada on aspects of the *NGSS*. Unfortunately, although several other small-scale projects have been conducted in the more populated regions of the state, most rural areas have basically been asked to implement reforms with little guidance. These individual efforts, most of which were run by the Nevada RPDP and state universities, fell short of achieving the adoption team's goal of reaching all teachers from across the state.

In the second year, a state grant allowed the two major Nevada universities and RPDPs from across the state to collaborate on training implementation specialists in each region of the state. In the third year, these specialists worked with the state transition team to provide professional development in each region of the state. However, despite these efforts, most of the teachers in the state have not received even awareness-level training on the *NGSS* in the third year, which is when the new standards-aligned science tests will be delivered. Clearly, the three-year adoption cycle was not adequate for training all teachers at even the basic level. Before any significant effects can be seen on teaching and learning, the process will need to go on for much longer.

Professional Development Through Lesson Study and PLCs

In exploring the *Framework* with cohorts of teachers, we developed an effective professional development model for supporting *NGSS* implementation. To begin, small groups of teachers attend a one- to three-day awareness workshop that introduces them to the *NGSS* and three-dimensional learning. Then, the teachers return to their schools and work with their teaching teams in PLCs. The PLC lesson-study design requires teachers to construct units of instruction for their specific grade levels. Teachers work together with *NGSS* mentors to unpack core ideas, identify anchor phenomena, write or select guiding questions, and develop phenomenon-based content sequences before collaborating on lesson plans that integrate the three dimensions of learning. Rather than stand alone, these plans must lead into a unit of instruction centered on phenomena. Teachers and mentors then work together to review the plans and provide one another with feedback. Then, they assemble individual plans into a unit of instruction.

During this review process, the participants use the EQuIP rubric from Achieve, Inc. (2016) to identify components that are necessary for *NGSS*-compatible instruction. The rubric requires teachers to look for evidence of three-dimensional learning, *NGSS*-aligned instructional supports, and methods of monitoring student progress that align with the goals of the new standards. Some groups go so far as to highlight the parts of their plan that focus on core ideas, science and engineering practices, and crosscutting concepts orange, blue, and green, respectively. Doing this helps teachers to determine whether or not a lesson incorporates a balance of the three dimensions of learning. In the EQuIP rubric's Category 2, teachers focus on adding relevance and authenticity by building progressions and utilizing students' prior knowledge.

After lessons are approved and implemented, teachers provide and receive additional feedback by observing one another's lessons and then discussing their observations. Category 3 of the rubric is used to measure three-dimensional student learning and guide formative assessments. This portion of the lesson study spans several months so that teachers have enough time to shift toward consistently providing three-dimensional instruction.

At the end of the process, the groups review student-learning products for evidence of the three dimensions of learning. Even after elaborate lesson planning and explicit three-dimensional teaching, it is common for student work to still require balance among the three dimensions.

Conclusion

California and Nevada adopted the *NGSS* one day apart. In this chapter, we have reviewed the two states' different transition strategies to identify similarities and differences. We found professional development, especially in the form of lesson study and mentoring, to be key for successfully implementing *NGSS*-aligned three-dimensional teaching and learning. Ultimately, the transition to full *NGSS* alignment will require much more time and resources than either California or Nevada was originally prepared to provide. Still, preliminary evidence shows that three-dimensional instruction increases student engagement with learning.

The *NGSS* promotes "science for all"—providing equitable access to scientific literacy for all students:

Successful application of science and engineering practices (e.g., Constructing Explanations, Engaging in Argument from Evidence) and understanding of how crosscutting concepts (e.g., Patterns, Structure and Function) play out across a range of disciplinary core ideas (e.g., Structure and Properties of Matter, Earth Materials and Systems) will demand increased cognitive expectations of all students. Making such connections has typically been expected only of "advanced," "gifted," or "honors" students. The *NGSS* are intended to provide a foundation for all students, including those who can and should surpass the *NGSS* performance expectations. At the same time, the *NGSS* make it clear that these increased expectations apply to those students who have traditionally struggled to demonstrate mastery even in the previous generation of less cognitively demanding standards. (NGSS Lead States, Appendix D)

Due to their diverse student populations, California and Nevada uniquely struggle with this aspect of *NGSS* implementation. In the past, limited access to science instruction for elementary students—often non-native speakers of English—at underperforming schools precluded any paths toward secondary education in science. Significant expansion in the use of inquiry-based science instruction over the last decade has improved matters considerably, transforming diversity from an obstacle to an opportunity. In Nevada and California, the concept of "science for all" will truly be tested, as will the great promise of the *NGSS* as a whole.

References

Achieve. 2016. *The EQuIP Rubric. www.nexgenerationscience. org/sites/default/resource/files/EQuIP.*

Bybee, R.W., J. A. Taylor, A. Gardener, P. Van Scotter, J. Carlson Powell, A. Westbrook, and N. Landes. 2006. *The BSCS 5E Instructional Model: Origins, effectiveness, and applications.* Colorado Springs, CO: BSCS.

California Department of Education. 2000. *Science content standards for California public schools, kindergarten through grade twelve.* Sacramento, CA: California Department of Education.

DiRanna, K., E. Osmundson, J. Topps, L. Barakos, M. Gearhart, K. Cerwin, D. Carnahan, and C. Strang. 2008. *Assessment-centered teaching: A reflective practice.* Thousand Oaks, CA: Corwin.

DiRanna, K., J. Topps, K. Cerwin, and S. Gomez-Zwiep. 2009. Teaching learning collaborative: A process for supporting professional learning communities. In *Professional learning communities for science teaching:*

Lessons from research and practice, ed. S. Mundry and K. E. Stiles, 34–54. Arlington, VA: NSTA Press.

Garet, M., S. Porter, L. Desimone, B. Birman, and K. Suk Yoon. 2001. What makes professional development effective? Results from a national sample of teachers. *American Educational Research Journal* 38 (4): 915–945.

Gomez Zwiep, S., W. J. Straits, K. R. Stone, D. D. Beltran, and L. Furtado. 2011. The integration of English language development and science instruction in elementary classrooms. *Journal of Science Teacher Education* 22 (8): 769–785.

Jeanpierre, B., K. Oberhauser, and C. Freeman. 2005. Characteristics of professional development that effect change in secondary science teachers' classroom practices. *Journal of Research in Science Teaching* 42 (6): 668–690.

National Governors Association Center for Best Practices and Council of Chief State School Officers (NGAC CCSSO). 2010. *Common Core State Standards in mathematics and language arts.* Washington, DC: NGAC and CCSSO.

National Research Council (NRC). 1996. *National Science Education Standards.* Washington, DC: National Academies Press.

National Research Council (NRC). 2015a. *Guide to implementing the Next Generation Science Standards.* Washington, DC: National Academies Press

National Research Council (NRC). 2015b. *Science teachers learning.* Washington, DC. National Academies Press.

NGSS Lead States. 2013. *Next Generation Science Standards: For states, by states.* Washington, DC: National Academies Press. *www.nextgenscience.org/ next-generation-science-standards.*

Reiser, B. 2013. What professional development strategies are needed for successful implementation of the Next Generation Science Standards? Paper presented at the Invitational Research Symposium on Science Assessment in Washington, DC.

Smith, A., E. Banilower, M. Nelson, and P. Smith. 2013. *The status of secondary education in the United States: Factors that predict practice.* Chapel Hill, NC: Horizon Research.

Trygstad, P., P. Smith, E. Banilower, and M. Nelson. 2013. *The status of elementary science education: Are we ready for the Next Generation Science Standards?* Chapel Hill, NC: Horizon Research.

Vallett, D., H. Deniz, K. Carroll, B. Sibley, and E. Adiblli. 2016. Implementing the NGSS: Results of statewide professional development project. Paper presented at the National Association of Research in Science Teaching International Conference, Baltimore.

David T. Crowther *is a professor of science education and executive director of the Raggio Research Center for STEM Education at the University of Nevada-Reno. At the time of this writing, he is also president-elect of NSTA. Crowther began his career as an elementary science teacher, then pursued graduate work in science education and biology. In a career spanning more than 25 years, he has authored numerous journal articles and national presentations, and he has secured more than 15 million dollars in external funding. He can be contacted by e-mail at crowther@unr.edu.*

Susan Gomez Zwiep *is a professor of science education at California State University-Long Beach and a regional director for the K–12 Alliance at WestEd. Before taking on these roles, she taught middle school science in the Los Angeles area. She has authored numerous journal articles and chapters related to science teaching, science curriculum, and professional development. Gomez Zwiep's work has been recognized by the Association of Science Teacher Education, which gave her their Innovation in Teaching Science Teachers Award, and by California State University at Long Beach, which bestowed on her the Early Academic Career Excellence Award.*

CHAPTER 8

Constructing Explanatory Arguments Based on Evidence Gathered While Investigating Natural Phenomena in a Professional Development Context

Kevin J. B. Anderson

In this chapter, we share examples from experiences that support teacher professional learning about scientific arguments based on evidence. We discuss foundational elements of engaging students in science talk and using the claims-evidence-reasoning framework for their conclusions, then delve into strategies to help teachers build formative and summative assessments and to engage students in modeling as part of their explanations. Throughout, we stress the importance of effective, cyclical professional development rather than one-time workshops.

Introduction

Small groups of sixth grade students at Replicon Middle School are discussing longitudinal data provided by the teacher on populations of two species of bats, as illustrated in Figure 8.1 (p. 70).

Ishmael: "Well one is going up and the other is going down."

Hannah: "It's not going up very much."

Sam: "Yeah, the other is going down a lot."

Hannah: "So, lots of the little brown bats are dying ... but we're getting more, uh, big brown bats."

Ms. Cooper (stopping by): "Can anyone say more about that? What's the specific evidence?"

Sam: "In the graph, these bats have gone from like 8,000 to 300, but these bats have gone from like 900 to 1,500."

Ishmael: "Um, why would some be dying but not the other ones?"

Ms. Cooper: "We're going to investigate that. I'd like your group to think of some ideas."

When the whole class reconvenes, Ms. Cooper leads a discussion about observations and questions that came up in the small-group conversations.

Ms. Cooper: "What did you observe in this data?"

Alicia: "Lots of bats are dying!"

Sam: "7,700 bats died. I did the math."

Hannah: "Yeah, but, there's more big brown bats now."

Ishmael: "Ms. Cooper, why are some bats dying?"

Ms. Cooper: "Who has an idea for an answer to Ishmael's question?"

Figure 8.1. New Jersey Summer Bat Count Longitudinal Data

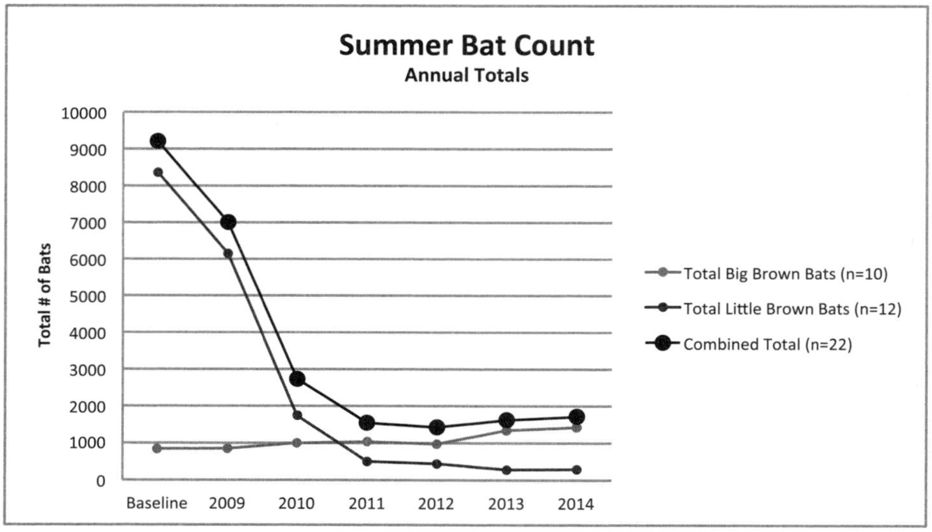

	Baseline	2009	2010	2011	2012	2013	% Change in # Bats
Total Big Brown Bats	839	850	998	1,034	971	1,339	59.6%
Total Little Brown Bats	8,361	6,153	1,734	508	451	286	-95.4%
Combined Total	9,200	7,003	2,731	1,542	1,423	1,625	-82.3%

Source: Conserve Wildlife Foundation of New Jersey 2016.

Note: n = 22 sites

If students are going to meaningfully engage in arguing with evidence, they need a phenomenon that enables them to investigate, make claims, and analyze related data. The same should be true of teacher groups engaged in professional development. Given data about a phenomenon, they generate a series of observations and questions that they share with other groups, then brainstorm possible causes and answers as well as ways of finding evidence for these.

To effectively deliver research-based science instruction aligned with the *Next Generation Science Standards* (*NGSS*; NGSS Lead States 2013), teachers must first experience it for themselves (Loucks-Horsley and Matsumoto 1999). Many teachers' foundational experiences with science learning did not reflect this type of instruction. Further, for such learning to have a lasting effect on instruction, it has to move beyond the one-day workshop and become embedded within teacher practice (Bredeson 2003). Although the strategies discussed in this chapter could be used for

short-term experiences, we intend for them to generate ongoing conversations among colleagues.

Science Talk Moves

Science talk moves are handy phrases that, when used over time, encourage students to listen to one another and more fully explain their thinking (TERC 2011). Because these moves are applicable across content areas, it is helpful to consider how they can be linked to specific science learning objectives. In a workshop, small groups of educators could each be assigned one move to discuss and tasked with creating a list of related science learning targets.

The science and engineering practices and crosscutting concepts of the *NGSS* can help to inspire learning targets. Table 8.1 shows a series of questions that can be used to deepen discussion initiated by science talk moves and support specific learning targets, thus enabling students to create high-quality scientific

Table 8.1. Talk Move Questions Based on the Science and Engineering Practice

Practices	Crosscutting Concepts
• What questions come up as you observe …? • What does this data tell us about …? • What did you notice when you …? • What if we changed this variable…? • What have you learned about … that's related to …? • Why did you choose this solution…? • What is your evidence that this is a credible source...?	• What patterns do you see …? • What do you think caused that to happen ...? • What would happen if we made it bigger (or smaller) ...? • Why did it stay the same (or change) …? • How does this structure help the … to ….?

explanations and build on their own work through science and engineering practices and crosscutting concepts.

During professional learning, educators can run through the prompts in Table 8.1 in a cycle—a technique that they can then continue to use in their own collaborative practice. In their first meeting, educators collaboratively read through and discuss the prompts and generate other examples in the process. Then, they decide on two or three of the newly developed questions to ask of groups or individual students multiple times throughout a lesson. During the next meeting, educators share their experiences using science talk moves in their classrooms and collaboratively reflect on the ways in which these prompts seemed to affect student responses and student learning.

Claims-Evidence-Reasoning (CER) Framework for Scientific Explanation

In the claims-evidence-reasoning (CER) framework, a claim is not necessarily the same as a hypothesis; it is a reasoned answer to the question being investigated, which students then back up with evidence and reasoning from their investigations and research (McNeill and Krajcik 2008). However, before teachers require students to write claims, they should first collaboratively do so themselves in a professional learning setting. Ideally, they can create a leveled progression of explanations based on claims, evidence, and reasoning to use as examples of distinct levels of student achievement. Alternatively, teachers could sort student papers from an already completed investigation into ability levels, then reflect on the factors that differentiate one level of ability from another.

Generally, while going through the process of writing CER explanations, a series of five steps guides the work. First, the main concepts from the current series of lessons are described, eliciting the scientific information most relevant to the reasoning. Second, the overarching question under investigation is identified. Third, a succinct claim is made that answers that question. Fourth, evidence derived from details of the investigations is used to support this claim. This may be given in a variety of forms, such as written descriptions, lists, tables, charts, or drawings. Fifth and finally, the main scientific concepts are used in the reasoning to show that the evidence supports the claim.

Generally, students and educators find reasoning to be the most difficult part of this process. For example, after learning about ecosystems, a student might claim that the introduction of an invasive species would reduce the number of native species. The student could provide evidence based on research that shows that populations of native fish have decreased the most in areas inhabited by Asian carp. The reasoning would then describe the scientific ideas of competition for limited resources and interdependent relationships in ecosystems, connecting the evidence to the claim.

Once teachers have written their own CER explanation for a current unit, they can help students, especially those new to the process, by replacing key ideas with blank lines to scaffold the process.

Choosing Among Explanations

In *Uncovering Students' Ideas in Astronomy*, Keeley and Sneider (2012) provide formative assessment probes that help students effectively develop explanations. The probe presents students with a phenomenon and asks them to decide among several possible explanations for it, explaining their choice. Morgan and Ansberry (forthcoming) suggest using the probe again later as a summative assessment. At the end of the lesson or unit, teachers ask students to select an explanation and provide evidence supporting it and invalidating the incorrect options.

Tools such as these serve as excellent means of assessment. However, because the specific topics and methods teachers use to engage their students may not align with them perfectly, I encourage teachers to create their own tools.

Here's an example. Teachers begin by watching a video of a series of temperature measurements of a beaker of ice water being heated on hot plate (or, if practical, they conduct the investigation for themselves). Asking the participants to share their observations prompts them to wonder why the temperature is not increasing. Then, the participants are asked to consider which of the following statements *best* explains the phenomenon:

1. The glass is very thick, so it insulates the water.

2. The heat energy transfers to the air.

3. The heat energy melts the ice.

4. The hot plate is not turned up enough.

After teachers discuss which is the proper answer and why, they speculate about reasons students might have for finding the other three answers plausible. It is important for all suggested explanations to be at least plausible, even though there is a single best answer. In the classroom, students would go on to further investigate their chosen explanation.

Following this discussion, teachers are asked to choose a phenomenon that is related to their content. If the teachers are unable to come up with their own idea, then one is assigned to them. Next, the groups work to develop their own set of explanations for their selected phenomenon. Teachers are also asked to identify the qualities of a good "distractor" explanation. I have found two resources to be especially helpful for developing these. The Vanderbilt Center for Teaching (Brame 2013) provides ideas for quality test-item alternatives that can easily be applied to this context. The center suggests that distractors should all be plausible, focused on the same content, and similar in length and form. They should

also connect to actual alternative conceptions that students may have. The American Association for the Advancement of Science (2016) provides an excellent resource for identifying alternative conceptions on their assessment website (see References).

Teachers often feel that if they simply explain an idea well, then students will fully understand a concept and reject all alternatives. However, research shows that students need to be guided through exploration and investigation of a phenomenon and arrive at the correct conclusions on their own for knowledge to stick (Committee on Undergraduate Science Education 1997).

Modeling to Support Effective Explanations

Whether explanations are written or spoken, it is important that they work in conjunction with scientific modeling. For example, rather than choosing among four written explanations of a phenomenon as described earlier, students might choose among pictures of four models and explain why their selection best illustrates the phenomenon. An even better option is for students to generate their own models based on evidence as part of their explanations for phenomena. During classroom discussions, students can use models to illustrate their ideas, and these models then go on to form an integral part of their CER explanations.

In a professional development setting, the process works as in the following example. Pairs of teachers receive two paper cups connected to each other at the bottom by a two-meter piece of yarn. Instead of being asked to explain this set-up, the teachers are instructed to explore it and to determine under which circumstances it does and does allow for the transmission of sound. Next, teachers are asked to draw a model of the phenomenon and to add graphics and text explaining the reasons that sound travels through the device in some situations but not in others. For example, they might draw one picture showing vibrations traveling from one cup to the other, and another showing the vibrations stopping at a hand holding onto the yarn.

After all the groups share their models and explanations, the teachers reflect on one another's models to determine the most important aspects. The activity concludes with a whole-group discussion of the following questions reflecting three aspects of modeling (Achieve 2016):

1. Which components are essential to the model? What does the explanation lose by not having certain components?

2. How are the components related? Is the relationship made explicit in the model through a time sequence, a comparison of two situations with a variable changed, directional arrows, or other such elements?

3. How does the model help explain the larger scientific phenomenon? In the example above, is there a clear representation of the sound-transmission process?

Connections to engineering help to engage students and extend their science learning. To illustrate this, teachers are next asked to reframe the investigation of the sound-transmission phenomenon as an engineering problem. The investigation could begin with a real-life scenario that provides a design challenge, such as the following:

> *You've been grounded from electronics after being caught texting in class again. As you storm to your room, you grab some random stuff from the kitchen, remembering something about sound traveling through different materials and a story from your grandmother about tin-can phones. Because you want to talk to your best friend next door without alerting anyone, you set out to design a communication device.*

Next, teachers use materials such as cups, string, tape, and paper clips to create a communication device themselves. As a group, the participants generate criteria and constraints, then create a drawing of their prototype. After some further collaborative investigation, teachers add explanations to the initial drawing detailing the process by which sound travels through the system. They also explain why sound travels better under some conditions than others (e.g., dry versus wet), which requires them to describe the science behind their design improvements using supporting evidence. The model can connect criteria for the engineering design to the scientific concept that sound transmission is affected by the material through which it travels.

Like students, most teachers need a lot of experience generating and revising their models before they

can excel at creating them. Many teachers continue to think of models as static diagrams that students re-create from a book or as tools from which students decipher information. In fact, as *A Framework for K–12 Science Education* (NRC 2012) makes clear, scientific modeling requires active sense-making and is an important component of arguing with evidence.

At the end of the professional development session, participants are asked to reflect on a phenomenon that students could model as part of their current unit and to find ways of reframing learning as solving an engineering problem. Teachers talk through these ideas in pairs and share them with the broader group for further feedback.

Building Assessments of Evidence-Based Argument

The strategies emphasized in this chapter—student discussion, claims-evidence-reasoning explanations, series of explanations that students evaluate, and scientific modeling—could all be used for the purposes of formative and summative assessment. First, however, teachers must engage in professional development to clarify the specific argumentation skill they want their students to exhibit. Do they want their students to simply share relevant evidence, or do they want them to justify how that evidence supports the argument? Do they want students to come up with and use the evidence themselves based on research or investigation, or do they just need to evaluate and draw conclusions from evidence provided to them?

In addition to determining criteria for assessment, teachers should collaboratively discuss the nuances of the skill that they will be assessing. A rubric can be useful for this purpose. Both the *NGSS* evidence statements (NGSS Lead States 2015) and Appendix F of the Standards offer details of possible sub-skills for a practice, and the appendix shows a possible developmental progression of those skills. In a professional development setting, teachers should first work to create a rubric before receiving examples. They begin by identifying one essential sub-skill, then determine a progression of abilities that they organize into a rubric. After the small groups have started work on individual documents, the teachers go through a virtual gallery walk to compare one another's rubrics. A sample rubric may also be provided for comparison. The teachers then share either an important aspect of

Table 8.2. Sample Rubric of the Sub-skill of Distinguishing Facts, Judgment, and Speculation

	1	2	3	4
Sub-skill: Distinguish between facts, judgment, and speculation	Student can generate a fact and an opinion and differentiate between the two.	Student differentiates between inferences, facts, and speculation.	Student can make sense of information to determine what is relevant. They can generate research-based inferences (claims).	Students can express reasoning about why their evidence or evidence found in research supports their inference (claim).
In This Example	Student can generate a factual statement about weathering and an opinion about why it's important to understand weathering.	Student can differentiate facts about weathering from inferences about possible causes of weathering or complete speculation about causes.	Students can read through graphical and textual information and pull out relevant evidence and information about freeze/thaw weathering. They can cite that information as evidence of freeze/thaw weathering.	Students can share their reasoning as to why particular evidence supports the claim that freeze/thaw weathering caused cracking in a road.

their rubrics or an idea for improvement gleaned from the virtual gallery walk.

Table 8.2 shows a fourth-grade sample rubric for evaluating students' abilities to argue with evidence. This example was developed for a lesson about the effects that freeze-thaw cycles have on road surfaces and focuses on the sub-skill of distinguishing among facts, judgments, and speculation.

During professional development, it is essential for "expert" thinking to be made transparent for participants. For this reason, facilitators may wish to make the steps of rubric creation explicit for educators at the outset of their work. If participants are already experienced with writing rubrics, a better idea is to provide the steps after they have completed an initial draft of their rubric, to facilitate reflection. For the example in Table 8.1, the steps are as follows:

1. Determine the specific skill that students are to demonstrate. Because erosion comes up in grade 4, the rubric was developed by referring to the section on the grades 3–5 band in Appendix F of the *NGSS*. Appendix F mentions the ability to distinguish among facts, reasoned judgment

based on research, and speculation. This skill was selected as the focus.

2. Determine the criteria for proficiency. For the rubric in Table 8.1, teachers reviewed evidence statements, particularly the sections noting that students are to "describe the evidence, data, and/ or models that support the claim" (NGSS Lead States 2015). To be considered proficient within the progression, students must work with data independently and make inferences; this becomes the third level of proficiency in the sample rubric.

3. Identify a starting point by reflecting on students' abilities. Many fourth-grade students can differentiate between a fact and an opinion after some coaching, so this skill was used as the baseline in the sample rubric. This baseline ability also aligned with the expected skills from the previous grade band, K–2, based on the progression of practice.

4. Write the second level of proficiency by describing a midpoint between the first and third levels. In the sample rubric, level three requires students to generate examples of facts, inferences, and speculation, so it makes sense that level two would ask

them to identify and differentiate among these types of information. Before describing the fourth level of proficiency, it can be helpful to consult the appendix and identify areas of increased rigor. In the sample rubric, abilities that exceed expectations are demonstrated when students show a deep level of understanding by specifying how their evidence supports an inference or claim.

When developing assessment rubrics, it is important to consider common shortcomings, such as the tendency of middle school students to simply dismiss evidence that is at odds with their initial conceptions (University of California Museum of Paleontology 2016). Extra support should be incorporated into classroom activities to help students overcome these challenges. Additionally, common shortcomings could be described in the appropriate rubric levels, not only to allow for a fair and honest evaluation, but also so that students can be aware of them and strive to surmount them.

Once the skill progression has been determined, formative assessments can take many forms. However, before sharing ideas developed for other classrooms, an educator group should first generate its own based either on the rubric creation process or on a method the teachers have previously employed.

In addition to determining a type of formative assessment, educators need to establish a plan for making use of the information the assessments generate. If most students are struggling, possible responses include providing more heavily scaffolded instruction in the next part of the lesson or conducting a demonstration of the desired skill but requiring the students to fill in key elements. For example, if students have trouble engaging in evidence-based argumentation, the teacher could prepare a handout that lists sentence starters or possible evidence. To complete the handout, students could be required to give the reasons that they chose to include or omit each piece of evidence available. These types of activities can be individualized according to student ability, and students who are already proficient can be allowed to complete the argument statement without the limitations of the scaffold. For additional reinforcement, the teacher can engage the whole class in a guided discussion that evaluates a series of arguments by noting whether they include relevant evidence and, if so, how the evidence supports each claim. Students can

then work in small groups, ideally with a mix of skill levels, to collaboratively revise their arguments.

If most students fall in the middle of the performance progression, the teacher might provide further time for group reflection and sharing. Ideally, the additional reflection should happen immediately, possibly even before all groups have presented their models. After the first few groups share, the whole class can discuss positive attributes as well as suggestions for refinement. At this point, groups can reconvene to improve their models based on the discussion before the presentations continue. The next time modeling occurs in a lesson, the teacher can repeat the present-discuss-revise process to provide support. For example, if only half the class knows how to write testable questions, the class discussion after the first few presentations might focus on which questions can be tested and which cannot, with students required to justify their conclusions. The groups then reconvene to revise their questions. Students who are still struggling after this can then work in a large group to talk through a present-discuss-revise process while the rest of the class starts work on independent research or brainstorming about how to design an investigation that will answer their testable question.

If most students can proficiently perform the task being assessed, the teacher should note which ones are still struggling and provide them with individualized support and feedback the next time they engage in a modeling activity. The teacher might also group these students together to more efficiently provide them with focused attention and perhaps additional scaffolds. Alternatively, struggling students could be paired with proficient students who can be trusted to meaningfully support their partner's learning rather than simply give away the answers. If any students continue to struggle after these remediation strategies, they can use the class-reviewed arguments to model their own revisions, with the teacher reviewing their work to ensure that they can back up their claims with evidence.

Once the participants in a professional development session are comfortable designing formative assessments and using them to inform instruction, they are asked to share in small groups and review their assessments with a particular science and engineering practice in mind, such as arguing with evidence. They analyze each assessment together, identifying tasks that require students to work with evidence to support an assertion. At times, a sample assessment may be provided if groups either lack a suitable example or if it seems they would be more comfortable critiquing a sample created by a neutral party; these can usually be easily located online.

As teachers work to create or revise their assessments, they should familiarize themselves with and make use of the helpful tools created by the Research and Practice Collaboratory (see *www. researchandpractice.org/wp-content/uploads/2016/02/ StepsToDesigningaThreeDimensionalAssessment_v4.pdf*). Two of these tools include a step-by-step guide for creating quality assessments; another integrates science and engineering practices into assessments that include specific supports for arguing with evidence.

Conclusion

Engaging teachers in active professional development opportunities is critical for improving practice. Educators must personally engage in science talk, crafting claims-evidence-reasoning arguments, comparing and analyzing possible explanations, creating scientific models to support explanations, and designing assessments and assessment processes before they employ these strategies in their classrooms. One suggested resource for this type of professional learning is the Council of State Science Supervisors' Science Professional Learning Standards (see *www.csss-science.org/downloads/SPLS.pdf*), which helps educators implement and sustain learning structures and evaluate the attendant learning.

Educators must habitually pause to reflect on their use of the methods they explore during professional development, and all evidence of student understanding should be cyclically reviewed by teacher teams. One effective model, based on the APEX model described by Thompson et al. (2009), is shown in Figure 8.2. In my experience, this type of cyclical work is essential for professional development, particularly when leaders act as coaches rather than as presenters or facilitators. A single-session workshop rarely results in changes to actual classroom practices.

As a final note, I especially enjoy collaborating on these learning cycles outdoors. Whether I am working at a school or in a conference center, I like to go on a walk around the community at some point in the day with participants to help them see how investigations and accrual of evidence can occur beyond the classroom walls. During these outings, participants ask questions

Figure 8.2. APEX Model of Collaborative Inquiry

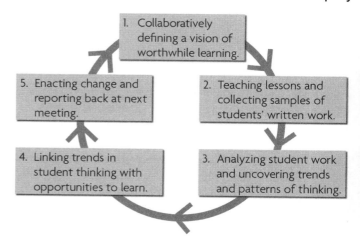

and consider possible phenomena to study. I make it a point to always pay close attention to these informal discussions, as they are just as necessary for professional learning as they are for classroom learning. If we want teachers to engage their students in learning activities centered on observation, inquiry, and investigation, then we must ensure that they have personal experience with the activities themselves.

References

American Association for the Advancement of Science (AAAS). 2016. AAAS science assessment. Washington, DC: AAAS. *www.assessment.aaas.org/topics*.

Brame, C. 2013. Writing good multiple choice test questions. *https://cft.vanderbilt.edu/guides-sub-pages/writing-good-multiple-choice-test-questions*.

Bredeson, P. V. 2003. *Designs for learning: A new architecture for professional development in schools.* Thousand Oaks, CA: Corwin.

Committee on Undergraduate Science Education. 1997. Misconceptions as barriers to understanding science. In *Science teaching reconsidered: A handbook,* ed. National Research Council. Washington, DC: National Academies Press.

Conserve Wildlife Foundation of New Jersey. 2016. Summer bat count. *www.conservewildlifenj.org/protecting/projects/bat/bat-count*.

Council of State Science Supervisors. 2015. Science Professional Learning Standards. *www.csss-science.org/downloads/SPLS.pdf*.

Janulaw, S. 2016. Know your students: Implications for student learning about the nature and process of science. *Understanding Science.* Berkeley, CA: University of California Museum of Paleontology. *www.undsci. berkeley.edu/teaching/68_implications2.php.*

Keeley, P., and C. I. Sneider. 2012. *Uncovering student ideas in astronomy: 45 new formative assessment probes.* Arlington, VA: NSTA Press.

Loucks-Horsley, S., and C. Matsumoto.1999. Research on professional development for teachers of mathematics and science: The state of the scene. *School Science and Mathematics* 99 (5): 258–271.

McNeill, K., and J. Krajcik. 2008. Inquiry and scientific explanations: Helping students use evidence and reasoning. *Science as inquiry in the secondary setting,* ed. J. Luft, R. Bell, and J. Gess-Newsome. Arlington, VA: NSTA Press. *www.katherinelmcneill. com/uploads/1/6/8/7/1687518/mcneillkrajcik_nsta_ inquiry_2008.pdf.*

Morgan, E., and K. Ansberry. 2017. *Picture-perfect STEM lessons, 3–5: Using children's books to inspire STEM learning.* Arlington, VA: NSTA Press.

NGSS Lead States. 2013. *Next Generation Science Standards: For states, by states.* Washington, DC: National Academies Press. *www.nextgenscience.org/ next-generation-science-standards.*

NGSS Lead States. 2015. Evidence statements. *www. nextgenscience.org/evidence-statements.*

TERC. 2011. *The inquiry project.* Washington, DC: TERC. *https://inquiryproject.terc.edu.*

Thompson, J., M. Braaten, M. Windschitl, B. Sjoberg, M. Jones, and K. Martinez. 2009. Examining student work: Evidence-based learning for students and teachers. *The Science Teacher* 76 (8): 48–52.

Kevin J. B. Anderson *is the science education consultant at the Wisconsin Department of Public Instruction. He also helps coordinate STEM Wisconsin, a statewide STEM education network. Previously, he worked as a regional STEM consultant, science and mathematics teacher, and education researcher. He received his PhD at the University of Wisconsin-Madison and his BA and MA at Stanford University. At work, he enjoys helping teachers understand and implement the NGSS. Most of all, he likes to explore the world with his family. Anderson can be reached by e-mail at* Kevin.Anderson@dpi.wi.gov.

CHAPTER 9

Examining Student Work and Interactions That Reflect the *NGSS*

Deb M. Kneser

In this chapter, we discuss the collaborative analysis of student work as an effective method of professional development. By engaging in a collaborative review of student work, science educators can analyze the learning experiences they have designed for their students and determine their effectiveness. We begin with an overview of the need for teacher professional development, followed by a discussion of protocols for collaboratively examining student work. We conclude with a short vignette of teachers engaging in this process.

Introduction

Research shows that good teachers can substantially improve student achievement. Highly effective instruction can result in student gains a full two months ahead of average gains. For example, a full year of learning happens in eight months as opposed to the average 10 months in a school year. Ineffective instruction can result in less than a full year of gains in a 10 month learning cycle and these effects are only compounded over time (DuFour and Marzano 2011; Hattie 2008; Wright, Horn, and Sanders 1997). The National Research Council (NRC 2015) offers the following guidelines for effective teacher professional development:

> Base design of professional learning on the best available evidence. When designing professional learning experiences, district and school leaders and providers of professional development should build on the key findings from research. Professional development should (1) be content specific;

(2) connect to teacher's own instructional practice; (3) model the instructional approach being learned and ask teachers to analyze examples of it; (4) enable reflective collaboration; and (5) be a sustained element of a comprehensive and continuing support system. For sustained implementation, research shows that principals' understanding of and support for instructional change is key. (p. 38)

Professional Learning

Quality of instruction is the school-related factor that can most improve student achievement (Hanushek et al. 2011; Nye, Konstantopoulos, and Hedges. 2004). When teachers receive quality professional development for an average of 49 hours spread over 6 to 12 months, they can increase student achievement by as much as 21 percentile points (Yoon et al. 2007). The current call to increase student achievement in science accentuates the need for effective science instruction. Effective science teachers establish rigorous and challenging learning environments that engage students in science. According to Harvard Graduate School Lecturer Victor Pereira, "educators should portray science as acquiring skills, rather than memorizing facts. If the classroom focuses on the scientific process of discovery, more students will be engaged in the subject matter" (Shafer 2015).

Science teachers need the opportunity to develop their practice. A National Staff Development Council report notes that, "while U.S. policy initiatives increasingly reflect an understanding that effective teaching and school leadership are critical to the quality of education that students receive, there is often less recognition that teacher professional development is a key element of school reform" (Wei et al. 2010, p. 1). The report goes on to note that much of the professional development available to teachers is disconnected from their practice. Connecting professional learning to teachers' practice must become a key element of school reform.

Professional Learning and Professional Learning Communities

A report by the Boston Consulting Group (2014) found that teachers prefer professional development that helps them plan and improve their instruction, is teacher-driven, includes hands-on strategies relevant to their classrooms, is sustained over time, and recognizes that teachers are professionals with valuable insights. These facets of professional learning can all be met successfully in a professional learning community (PLC), defined by DuFour and Reeves (2015) as an "ongoing process in which educators work collaboratively in recursive cycles of collective inquiry and action research in order to achieve better results for the students they serve."

DuFour and Reeves (2016) explain that when teachers work collaboratively in a true PLC, the PLC should be driven by these four questions:

- What do we want students to learn?
- How will we know if they have learned it?
- What will we do if they have not learned it?
- How will we provide extended learning opportunities for students who have mastered the content?

Analyzing student work in a PLC and using these four questions focuses the dialogue on student learning. In a focused PLC, collaborative teams of educators use evidence of student learning as a basis for collective inquiry into their own instructional practice. Focusing teacher dialogue on real evidence of learning (or lack thereof) can bring about true change in instruction. We cannot simply keep repeating more of the same ineffective lessons and demonstrations if our goal is to improve student achievement.

Why Analyze Student Work?

Engaging in a collaborative process of looking at student work in science allows a group of educators to analyze the learning experiences they have designed for their students and determine their effectiveness. The benefits of teachers working together have been documented in multiple studies (Ball and Cohen 1999; Hargreaves and Fullan 2012). The Carnegie Corporation of New York's report (2009) explicitly called for science teachers to work collaboratively to change instruction. When teachers collaboratively analyze student work, they can build understanding and agreement about effective instruction and curriculum implementation (Allen 2013). Full implementation of the *NGSS* requires teachers to work together and learn from each other. The process encourages teachers to consider the following questions. For the required core ideas, crosscutting concepts, and practices,

- Which strengths do students demonstrate?
- What learning gaps are apparent?
- How do we know students have learned what they need to?
- What do we do when students are struggling or already proficient with the *NGSS*?

Kazemi and Franke (2004) studied monthly meetings of a group of teachers participating in a PLC focused on strategies that their students used to solve math problems. These meetings resulted in shifts in the teachers' thinking about student learning. The teachers paid closer attention to details of students' thought processes, and they reflected on multiple learning trajectories for students. Windschitl, Thompson, and Braaten (2011) came to similar conclusions, adding that improvements occurred primarily among teachers who viewed instructional challenges as "puzzles of practice" to be investigated with colleagues. Rather than view struggles in the classroom as rooted in "problems with students," these teachers explored all plausible causes for their underachievement.

According to Langer, Cotton, and Goff (2003), "the most important benefit of collaboratively analyzing student learning is that students learn more." Additional benefits that I have witnessed include the following:

- Increased collaboration and sharing among educators
- A shared vision for curriculum implementation
- Increased professional dialogue and a shared knowledge about curriculum, assessment, and students
- Increased clarity about intended outcomes
- A shared vision of student success
- Fostering a culture that collaboratively assesses the quality and rigor of teacher work
- An increased feeling among educators that they are "all in this together."

Leadership Roles in Analyzing Student Work

Principals, curriculum directors, department chairs, and teacher leaders are instrumental in supporting teachers who are working to improve their practice by analyzing student work in PLCs. Leithwood and colleagues (2004) found that classroom instruction is the only factor with a greater effect on student achievement than school leadership. For the work of PLCs to be successful, leaders must be invested in and support the process with vision, time, and resources. Support from school leaders will ensure continual professional learning, a positive school climate, and success for all students (Wallace Foundation 2012).

Leadership is instrumental in establishing and maintaining a healthy school culture and a positive science teaching and learning environment. The responsibility to teach science well is collective. As Sato, Bartiromo, and Elko note, "Ultimately, student scientists must learn science in a culture that emits the values, attitudes, and beliefs related to the science we want them to culturally live and breathe" (Sato, Bartiromo, and Elko. 2016).

Implementing the *NGSS* requires an "all hands on deck" mentality. Teachers used to working independently within their own classrooms must shift perspectives, and the sense of community provided by a PLC can help them to do this. School and district leaders must understand that PLCs are essential for building the sense of community necessary to successfully adopt the *NGSS*. Reviewing student work and problem solving collaboratively moves teachers away from the isolating concept of "my students" and toward the community concept of "our students."

Using Protocols to Look at Student Work

In my work with teachers, I have used a variety of protocols based on work from the National Reform School Faculty (NRSF), which offers over 200 free protocols at its website (see *www.schoolreforminitiative.org/protocols*). Protocols are processes that guide dialogue among groups. Usually they are composed of a facilitated set of steps that allow participants to discuss and listen with deep reflection. Protocols structure discussions for predetermined purposes by providing feedback to colleagues about tasks they have designed for students, investigating a problem of instruction, or analyzing student work samples, often in relation to learning goals or curriculum standards (Easton 2009; Little, Gearhart, Curry, and Kafka 2003). With the introduction of *NGSS* and the three dimensions of learning, it is important that educators understand the levels at which their students are performing in all areas. Protocols can help them to glean information about curricular decisions and fidelity of implementation of the *NGSS*.

"For many educators, protocol-guided discussion of student work is an unfamiliar and not entirely comfortable way of working with colleagues" (Blythe, Allen, and Powell 2015). Once participants have become familiar with the norms and procedures for analyzing student work, they are ready to start. Typically, participants begin by posing a question that they hope to answer through close, collaborative examination of student work. This helps them to develop a shared sense of purpose and commitment. Examples of questions that PLCs often use during a review of student work include the following:

- What are students' strengths with regard to the required crosscutting concepts, disciplinary core ideas, or practices in the *NGSS*?

- What are the students' learning needs in science? Does the work they are doing build on these needs?

- Do students exhibit gaps in their learning of crosscutting concepts, disciplinary core ideas, or practices in the *NGSS*?

- What evidence is there that gaps exist in the implementation of the *NGSS*?

- How is student learning aligned to the *NGSS*?

After developing these questions, groups develop criteria to consider in their review. Participants are asked to bring forward examples of student work that fit the determined criteria. Collaboratively, participants examine the work and provide feedback to their colleagues, who are given time to reflect on the input. At the end, a final discussion is held to reflect on both student and teacher learning.

Analyzing Student Work

The practice of engaging students in scientific modeling features prominently in the *NGSS*, and *A Framework for K–12 Science Education* (*Framework*; NRC 2012) emphasizes the need for students to evaluate ideas to reach the best explanation. The *Framework* requires students to (1) know how to use primary or secondary scientific evidence and models to support or refute an explanatory account of a phenomenon and (2) identify gaps or weaknesses in explanatory accounts (their own or those of others).

Because building models and constructing explanations is so important for student learning, one school district in the Midwest decided to assess how much these skills were evident in its students' work. The district was trying to move away from having students simply present facts and definitions in their work and toward constructing explanations based on evidence.

The district used an existing PLC as the starting framework for examining student work. The team consisted of 11 middle school science teachers, a science teacher instructional coach, and the district curriculum coordinator. The middle school science teachers were asked to identify two assignments they had used that demonstrated building models and constructing explanations and to bring at least two samples of student work from each assignment to the PLC for analysis.

The student work reflected a variety of topics and content. Models explicated such disparate phenomena as the phases of the Moon, solar and lunar eclipses, plant transpiration, gravity, and osmosis. Each science teacher took a few minutes to discuss the work they'd brought in and shared the assignment for which it was created. Participants then used sticky notes to provide presenters with feedback and questions about their artifacts. In this structured, facilitated discussion model, teachers offered both "warm" (supportive) and "cool" (critical) feedback to their peers, who then reflected on the feedback with one another for about 45 minutes.

Next, participants worked together to review both the student work and their own instructional strategies, looking for patterns in the work and noting anything that surprised them. The discussion that followed was guided by the following four key questions identified by DuFour and Reeves (2016):

1. What do we want students to do with models and construction of explanations?

2. What do we consider to be evidence of students using models and constructing explanations in their work?

3. What are the next steps to facilitating a deeper learning about the practices of building models and constructing explanations?

4. How will we provide extended learning opportunities for students who have mastered the practices of building models and constructing explanations?

During the discussion, teachers further developed ideas for using models in their classrooms and identified examples of appropriate rigor for student assignments. Several teachers concluded that their own assignments for building models and explanations needed more rigor or authenticity. For example, one teacher had students draw the phases of the Moon from a diagram that the students had found on the internet. The students only had to replicate the original drawing, not construct their own model of the phases of the Moon. The student explanations were weak and basically just explained what the drawing demonstrated. Another teacher had students watch a video on the phases of the Moon, and then construct their own models from the information in the video. These students' explanations were more detailed and referred to the original models to support their claims.

Participants also discussed the value of "seeing into student work" to better understand the students' thought processes. They were surprised by how often they couldn't identify the student thinking that informed the work. When student thinking is visible, teachers are better able to uncover misconceptions, prior knowledge, reasoning ability, and degrees of understanding. This process led teachers to discuss classroom strategies they might use to help identify student thinking. The teachers asked the curriculum director to research such strategies and provide them with related professional development opportunities.

Several teachers asked for help rewriting their tasks from colleagues, having discovered that they were not challenging enough or otherwise poorly designed. Some of them noted that tasks integrating instruction with other subject areas or connecting it to real-world experiences led to increased student learning. The biggest discovery of the day was that all students needed help using evidence to construct explanations and recording their thinking more clearly.

When educators review student work on their own, they often do so through the narrow lens of the given assignment. Reviewing student work with colleagues offers a broader perspective that can yield greater insight about student learning and quality of instruction.

Closing Comments

Educators need to move beyond working in isolation and toward more professional collegiality. Using professional learning communities to analyze student work can help teachers make this transition. Protocols for analysis allow educators to engage in structured and meaningful conversations about teaching and learning. As teachers reflect on and discuss the complexities of instruction, they develop a common understanding of the content they expect students to learn and a common vision of the skills they want students to demonstrate.

Of course, we must always remember that it is not the protocols but the deep conversations and reflections we have about them that influence student learning.

References

Allen, D. 2013. *Powerful teacher learning: What the theatre arts teach about collaboration.* Lanham, MD: Rowman & Littlefield Education.

Ball, D. L., and D. K. Cohen. 1999. Developing practice, developing practitioners: Toward a practice-based theory of professional education. In *Teaching as the learning profession: Handbook of policy and practice,* ed. G. Sykes and L. Darling-Hammond, 3–32. San Francisco: Jossey-Bass.

Blythe, T., D. Allen, and B. S. Powell. 2015. *Looking together at student work.* New York: Teachers College Press.

Boston Consulting Group. 2014. *Teachers know best: Teachers' views on professional development.* Washington, DC: Bill & Melinda Gates Foundation. *https://s3.amazonaws. com/edtech-production/reports/Gates-PDMarketResearch-Dec5.pdf.*

DuFour, R., and R. J. Marzano. 2011. *Leaders of learning: How district, school, and classroom leaders improve student achievement.* Bloomington, IN: Solution Tree Press.

DuFour, R. and D. Reeves. 2015. Professional learning communities still work (if done right). *Education Week.* *www.edweek.org/tm/articles/2015/10/02/professional-learning-communities-still-work-if-done.html.*

DuFour, R., and D. Reeves. 2016. The futility of PLC lite. *Phi Delta Kappan* 97(6): 69–71.

Easton, L. B. 2009. *Protocols for professional learning.* Alexandria, VA: ASCD.

Hanushek, E. A. 2011. The economic value of higher teacher quality. *Economics of Education Review* 30 (3): 466–479.

Hargreaves, A., and M. Fullan. 2012. *Professional capital: Transforming teaching in every school.* New York: Teachers College Press.

Hattie, J. 2008. *Visible learning: A synthesis of meta-analyses relating to achievement.* London: Routledge.

Kazemi, E., and M. L. Franke. 2004. Teacher learning in mathematics: Using student work to promote collective inquiry. *Journal of Mathematics Teacher Education* 7 (3): 203–235.

Langer, G. M., A. B. Colton, and L. S. Goff. 2003. *Collaborative analysis of student work: Improving teaching and learning.* Alexandria, VA: ASCD.

Little, J. W., M. Gearhart, M. Curry, and J. Kafka. 2003. Looking at student work for teacher learning, teacher community, and school reform. *Phi Delta Kappan* 83 (5): 184–192.

National Research Council (NRC). 2015. *Guide to implementing the next generation science standards.* Washington, DC: National Academies Press.

National Research Council (NRC). 2012. *A framework for K–12 science education: Practices, crosscutting concepts, and core ideas.* Washington, DC: National Academies Press.

NGSS Lead States. 2013. *Next Generation Science Standards: For states, by states.* Washington, DC: National Academies Press. *www.nextgenscience.org/next-generation-science-standards.*

Nye, B., S. Konstantopoulos, and L.V. Hedges. 2004. How large are teacher effects? *Educational Evaluation and Policy Analysis* 26: 237–257.

Sato, M., M. Bartiromo, and S. Elko. 2016. Investigating your school's science teaching and learning culture. *Phi Delta Kappan* 97 (6): 42–47.

Shafer, L. 2015. Why science? Amid STEM enthusiasm, stepping back to consider the broader purpose of teaching and learning science. *Usable Knowledge.* Harvard Graduate School of Education. *www.gse.harvard.edu/news/uk/15/11/why-science.*

Wei, R. C., L. Darling-Hammond, and F. Adamson. 2010. *Professional development in the United States: Trends and challenges.* Dallas: National Staff Development Council.

Windschitl, M., J. Thompson, and M. Braaten. 2011. Ambitious pedagogy by novice teachers: Who benefits from tool-supported collaborative inquiry into practice and why? *Teachers College Record* 113 (7): 1311–1360.

Wright, S. P., S. P. Horn, and W.L. Sanders. 1997. Teacher and classroom context effects on student achievement: Implications for teacher education. *Journal of Personnel Evaluation in Education*, 11, 57–67.

Yoon, K. S., T. Duncan, S. W. Y Lee, B. Scarloss, and K. Shapley, K. 2007. Reviewing the evidence on how teacher professional development affects student achievement *Issues & Answers.* REL 2007–No. 033. Washington, DC: U.S. Department of Education, Regional Educational Laboratory Southwest.

Deb M. Kneser *is an assistant professor in the School of Education and the director of the Institute of Professional Development at Marian University in Fond du Lac, Wisconsin. Previously, she worked with Cooperative Educational Service Agency 6 in Oshkosh, Wisconsin, where she served as an educational consultant in curriculum, assessment, and instruction. Kneser is a former elementary classroom teacher with National Board certification, and she has extensive experience providing professional development sessions and presentations throughout the United States and internationally. She can be contacted via e-mail at dmkneser65@marianuniversity.edu.*

3

Teacher Preparation Courses for Preservice Science Teachers

CHAPTER 10

State Policies and Regulations for Implementing the *NGSS*

Stephen L. Pruitt

In this chapter, we address state policies and regulations that affect the implementation of the *Next Generation Science Standards* (*NGSS*; NGSS Lead States 2013). Though our focus is on the new standards, we also consider state and district policies and discuss how they can support or hinder *NGSS* implementation. Additionally, we consider the effect that federal policies, especially the Every Student Succeeds Act (ESSA), have on the perception and quality of implementation. Throughout this chapter, we explore issues of assessment, funding, instructional materials, and graduation policy.

Introduction

The components necessary to successfully implement the *NGSS* include leadership, educator and community involvement, and knowledge of three-dimensional learning. For all students to benefit from the *NGSS*, policies and regulations at the state and district levels must support quality implementation. Though no policies or regulations are able to fully enforce quality, they can set the conditions and direction of change, driving the conversation. Still, at the end of the day, educators will always be the ones selecting the types of activities they use in the classroom.

Too often, one crucial group is omitted from policy discussions: principals. If principals aren't cognizant of what it takes to successfully implement reforms, including the time necessary to devote to it, challenges will undoubtedly arise. Teachers need time to review the standards, modify lessons, and align assessments to the new standards. It is not practical or beneficial to implement change of this scope in a single year, much less over a single summer. Only once principals understand the magnitude of the work that must be done will they be able to foster a professional environment in which teachers can implement change at a reasonable, steady pace.

The sooner we are able to implement the *NGSS* and the principles of three-dimensional learning, the sooner we will have improved science education for our nation's children. New ideas and methodologies in education often seem to be implemented at a glacial pace, even when there is overwhelming evidence to support their efficacy. However, I have seen the *NGSS* draw incredible support and commitment from creative and motivated educators who are willing to invest their time and energy not because of policy or pressure, but because it is best for students. I urge all states and districts to ensure that our nation's educators can experience the benefits of the *NGSS* for themselves.

Over the past few years, I have seen many teachers implement the *NGSS* without even waiting for state adoption. Why would overworked teachers take it upon themselves to rewrite curricula, plan new lessons, and develop new assessments before this work has been mandated? The answer is simple: The research shows that the *NGSS* makes learning meaningful for students and aids retention, even on standardized tests that focus largely on simple recall. I urge all policymakers to join these forward-thinking educators on the front lines of developing curriculum and assessments at the state, district, and school levels. At the same time, let us never forget that the most powerful agent of change at all levels of education will always be the educator.

Federal Policy

The No Child Left Behind Act (NCLB) has been the blueprint for federal education policy for much of the careers of many current teachers. Though the policy did some good—by shifting the focus away from overall scores, for example, and by mandating equity for all students—the policy's focus on testing, on mathematics and reading, and on school accountability systems have also hurt our students in many ways. Don't get me wrong; I am a believer in testing. It is an objective measure of the opportunities given to students. But when we focus on it at the expense of other matters, we prevent students from receiving a well-rounded education. Some critics of the NCLB believe that it actually resulted in a decrease of science learning (Griffith and Scharmann (2008) especially at the K–5 level, and there may be some truth to this, though one could also argue that science hasn't been a priority in a lot of our schools for a long time.

Since its passage in 2001, NCLB changes have begun at the top and forced their way down to the states, districts, schools, and finally classrooms. Thankfully, ESSA, which is a new iteration of the law, grants states and districts greater accountability than before. Although this is exciting, it also entails a huge responsibility. For an accountability system to be fair and effective, all educators should be able to have input into its design and help guide its priorities.

Educators and other policymakers must bear in mind that not everything of value can be accounted for in a system, lest it cave under its own weight. But we do need to ensure that our students fully understand the basic tenets of science, and ESSA requires us to assess this. Simply adding science to standardized tests is not sufficient; states have been required to have science standards and assessments since 2005, yet we have not seen any great improvement in science scores or even much of a shift toward more science education. States, districts, schools, and educators must all do more in this regard.

State-Level Policies

In my experience, implementation of state education policies is most successful when the state includes shareholders in planning and develops the right metrics for determining compliance. Traditionally, state agencies have assessed the success of implementation by recording the number of teachers trained in the new policy, but this method has been shown time and again to be insufficient. Simply requiring teachers to complete a training program does not guarantee that they will incorporate what they learn into classroom activities. Rather, states need to take the extra step of conducting observations and communicating with both teachers and students through forums or surveys.

It is critical for educators to understand that measurements of implementation are not the same as evaluations of schools or teachers. The purpose of the observations and surveys or forums is strictly to evaluate state progress toward full implementation. In Kentucky, for example, the state department of education evaluates student work related to specific tasks not to "grade" it but rather to determine if the quality is improving over time.

Implementation is the most difficult component of policy change, because we often focus so entirely on simply enacting new standards that we forget about our end goal: to improve instruction. Policy is not about training people, developing assessments, or even communicating. It is about helping students learn. To this end, educators must focus on the following areas during implementation.

Adoption

Already, 17 states and the District of Columbia have adopted the *NGSS*. They have been proactive about planning implementation and engaging shareholders. Teachers cannot simply review the *NGSS* among themselves; they must include post-secondary faculty, legislators, governors' offices, parents, advocacy groups, and business and industry representatives in the discussion. Although restricting the adoption team to a small group makes reaching a consensus much easier, overlooking the potential contributions of other shareholders can sharply limit the effectiveness of science education outside the classrooms. It is our responsibility as educators to prepare students to contribute to society, and individuals who are contributing already are best able to tell us what students will need to know to do this successfully.

States should consider the context and climate during the adoption of the *NGSS* and select a time line that is appropriate for all districts. If individual teachers are prepared to move forward earlier, they can begin to implement the vision of the *Framework* in their own classrooms while still working under the old standards. I have been asked if the slow pace of adoption disappoints me, and the short answer is no. From the beginning, I have counseled states to take the time to get implementation right.

Assessment

According to *Developing Assessments for the NGSS* (NRC 2014), quality science assessments must be designed and delivered at both the classroom and state levels, and both must be equally valued. If we focusing exclusively on state-mandated high-stakes tests, we fail to empower our teachers to use formative assessments to inform their instruction and adapt activities to fit the needs of their specific students.

Though assessment literacy is critical to effective teaching, it is often overlooked during teacher education. In Kentucky, we are designing a system that engages teachers and values classroom assessment. Our goal is to promote good instructional practices that prepare students for success on an *NGSS*-aligned state-level assessment. Due to both cost and logistics, states cannot regularly administer performance-based assessment that utilize specialized materials and laboratory techniques, making local assessment vital to accurately assess *NGSS* implementation. At the same time, the state agency remains responsible for developing quality assessments aligned to the new standards that consist of more than multiple choice questions, require more than rote memorization, and are relevant in real-world contexts.

State agencies and educators must work together to develop assessments specifically designed to reveal three-dimensional learning, and states must at the very least pay attention to grades and subjects not assessed at the state level. In Kentucky, as in many states, we assess the minimum required by the federal government. Multiple factors prevent the addition of more tests, and frankly I would not want to increase the number of assessments to which students are subjected. Instead, our schools will submit student work from each grade, K through 12, to be evaluated in terms of quality and *NGSS* alignment. (Some shareholders are surprised that the state begins tracking science instruction as early as grades K–5, but these levels provide the base of knowledge and skills that students need to succeed in middle and high school, and thus important to assess.)

Graduation Policy

Currently, most states require that students complete three years of science to graduate from high school. Often, these requirements lack any further detail, although some states do specifically require students to learn biology, either directly through legislation or by requiring completion of a biology test. Some states further specify that students must study chemistry or physics, but few states require Earth sciences, which I feel is a serious oversight. These requirements are a vestige of a decision made in 1892 by the Committee of Ten (NEA 1894) to separate the sciences into the current subjects. I take issue with the current requirements, because few modern careers require their professionals to be knowledgeable about just one branch of science. Furthermore, one biology course or test is not sufficient to indicate scientific literacy. For that matter, having any number of science courses listed on a transcript does not demonstrate scientific literacy. Although the public may still not be ready to determine scientific literacy using the *NGSS* rather than through the more familiar course-based method, I hope the idea will continue to gain traction in other states as it is here in Kentucky.

Funding

No one likes to talk about funding, but everybody needs it. Federal funding from each of the Title programs provides opportunities for all schools, and states should ensure that their districts and schools are familiar with their options. Although state agencies cannot dictate how schools use the funds they receive, they should offer guidance for doing so effectively.

Other State-Level Considerations

I understand that compliance will always be an issue states, but the quality of educational experiences is what matters most. Since the NCLB was first implemented, states have been pushed by the United States Department of Education to complete assessments quickly if they want continued funding. This pressure led to many states rushing implementation without due fidelity. As an education community, we must adopt a mindset of quality and refuse to accept mere compliance.

District-Level Policies

In many states, districts hold much of the authority regarding instruction, but they still need the state to take the lead on policies (especially on controversial issues such as graduation requirements). Still, districts need not wait for state adoption before moving forward with the *NGSS*, and many haven't. When adopting new standards, both states and districts will have to convince the public that they're necessary. Aligning teams and communicating with parents is critical at this stage.

As previously noted, successful implementation requires a quality mindset, whether at the state or district level. For example, we can no longer simply trust vendors when they claim their product provides quality learning experiences. Policies and procedures need to be in place guaranteeing it. Districts also need to establish policies around instructional materials, assessment, professional development, and teacher leadership. Traditionally, instructional materials are presented to teachers for review, discussion is hurried, and there is little thought given to quality as defined by the standards. I would like to see policies requiring all educational products to be fully aligned to standards.

Conclusion

Although policy and regulations are critical drivers of change, they are far from the only ones. It is urgent that educators at all levels, business representatives, parents, and all other shareholders create a common vision for improvement. No amount of state-mandated regulations can force anyone to care about science or value the development of a scientific mindset. Regulations cannot fully ensure fidelity of implementation, and often cannot even dictate how funds are dispersed or used.

Rather than holding off until agencies within the federal or state government arrive at decisions regarding reforms, educators should embrace change for themselves. The time has come to stop doing only what we think we have a right to do and start doing what is right—and it is right to ensure that all students have access to a quality science education.

References

Griffith, G., and L. Scharmann. 2008. Initial impacts of No Child Left Behind. *Journal of Elementary School Science* 20 (3): 35–49.

National Education Association (NEA). 1894. *Report of the Committee of Ten on secondary school studies*. Chicago: American Book Company.

National Research Council (NRC). 2014. *Developing assessments for the Next Generation Science Standards*. Washington, DC: National Academies Press.

NGSS Lead States. 2013. *Next Generation Science Standards: For states, by states*. Washington, DC: National Academies Press. *www.nextgenscience.org/next-generation-science-standards*.

Stephen L. Pruitt *is the commissioner of education for the Commonwealth of Kentucky. Before accepting this position, he was senior vice president for Achieve, where he led the development of the NGSS. Commissioner Pruitt has authored several journal articles, chapters, and books, and he has held several state-level education policy positions. However, he has always been most proud of his time as a high school chemistry teacher in Georgia. Pruitt has received NSTA's Distinguished Service to Science Education Award as well as a lifetime membership to the Council of State Science Supervisors. He can be contacted by e-mail at stephenlpruitt@gmail.com.*

CHAPTER 11

Helping Prospective Teachers Understand Disciplinary Core Ideas and Crosscutting Concepts in an Elementary Education Methods Course

Norman G. Lederman, Judith S. Lederman, and Selina L. Bartels

The *Next Generation Science Standards* (*NGSS*; NGSS Lead States 2013) are an ambitious attempt to transform science learning. In addition to reframing the knowledge and skills expected of students, these revised standards address the methods through which science is to be taught and must be reflected in teacher preparation programs, including those for primary teachers, who set the groundwork for secondary instruction. In this chapter, we review the *NGSS* and then offer examples of elementary science methods teachers can use to apply the *NGSS* in the classroom.

Introduction

The *NGSS* promise to provide K–12 with an enriching, engaging, and comprehensive science education. The *NGSS* are more comprehensive than previous reform documents, such as the Benchmarks for Science Literacy (AAAS 1993) and the National Science Education Standards (NRC 1996). The *NGSS* promote the application of science and engineering practices, disciplinary core ideas, and crosscutting concepts together in the form of three-dimensional learning. The science and engineering practices and crosscutting concepts promote a deep understanding of the nature of science and comprise knowledge from the various science disciplines (e.g., physical sciences, life science, Earth and space sciences) along with overarching science themes (e.g., patterns; cause and effect; scale,

proportion, and quantity; systems and system models; energy and matter; structure and function; stability and change).

When addressing the challenges posed by implementation, it is important to bear in mind that the application of science and engineering practices is distinct from the process of inquiry. The precise definition of inquiry has long been a source of confusion among science educators, resulting in inquiry-based classrooms that train students to carry out the steps of a scientific procedure without understanding the logic behind those steps or even the reasons for performing the procedure at all. This tends to happen when the application of science skills is taught without sufficiently emphasizing the underlying science concepts. The *NGSS* attempt to rectify this issue by drawing connections among all three dimensions of learning. The specific practices of science and engineering addressed by the *NGSS* are as follows:

- Asking questions and defining problems
- Developing and using models
- Planning and carrying out investigations
- Analyzing and interpreting data
- Using mathematics and computational thinking
- Constructing explanations and designing solutions
- Engaging in argument from evidence
- Obtaining, evaluating, and communicating information

Notice that the practices all begin with a verb, identifying them as actions students will learn to take in the classrooms.

The science and engineering practices and the crosscutting concepts of the *NGSS* extend the following ideas integral to the nature of science:

- Scientific investigations use a variety of methods.
- Scientific knowledge is based on empirical evidence.
- Scientific knowledge is open to revision in light of new evidence.
- Science models, laws, mechanisms, and theories explain natural phenomena.
- Science is a way of knowing.
- Scientific knowledge assumes an order and consistency in natural systems.
- Science is a human endeavor.
- Science addresses questions about the natural and material world.

In contrast to the science and engineering practices, the extensions to nature of science do not begin with a verb. These are cognitive understandings, not actions. However, it is through performance of the practices that students most effectively develop their understanding of the nature of science (Bybee 2013; Lederman and Lederman 2014). Indeed, it can be argued that this understanding is more important than the ability to successfully perform the practices. After all, when confronted with personal or societal issues that are scientifically based, people don't usually run outside to perform an investigation; rather, they use the process through which

scientific knowledge is developed to make their decisions (Lederman and Lederman 2016). Because teaching scientific understanding aligned to the *NGSS* is a daunting task, especially for teachers at the elementary levels who typically possess a less comprehensive knowledge of science than their secondary peers, elementary teachers will need to participate in professional development and preservice teachers will need to learn about the *NGSS* in their preparation programs.

It is critical for educators to understand that they will not be able to address the standards in all three dimensions along with nature-of-science ideas in every single lesson. Concepts and skills addressed will depend on the subject matter, student readiness, and the context of the lesson in the year's curriculum. Rather than overwhelm both teachers and their students by trying to focus on every aspect of every standard all at once, it makes sense to address all grade-appropriate standards several times throughout the year and all three dimensions at various points throughout the learning sequence.

The following activities have all been successfully field-tested in elementary school classrooms, and, as such, are ideal for inclusion in a science methods course for elementary teachers. For each activity, the key pedagogical understandings, the targeted *NGSS* dimensions, and the nature-of-science extensions are clearly identified, and alternative foci and modifications are also provided.

Modeling *NGSS* in Elementary-Education Methods Courses

The *NGSS* has a separate set of standards for each K–5 grade. This is a distinct departure from the National Science Education Standards, which had one set of standards for the K–4 grade band and a separate set of standards for the 5–8 grade band. Another feature new to the *NGSS* is the inclusion of progressions for disciplinary core ideas across grade levels. Because preservice teachers do not know which grades they might be hired to teach, they must be taught lessons that are appropriate for each primary grade level as well as how to modify lessons for different grades.

Engineering practices and connections to the *Common Core State Standards* (*CCSS*; NGAC and CCSSO 2010) for mathematics and language arts must also be addressed in elementary teacher preparation programs. Teaching engineering practices can be especially challenging for science teachers, few of whom are engineers or have taken even a single engineering course, let alone taught engineering to elementary students. Nonetheless, we must now rise to this challenge and model the instruction of engineering alongside science content.

Integrating science instruction with mathematics and English language arts standards is usually less of a challenge. Most elementary teachers are responsible for teaching across the core subject areas, so they're at least a little familiar with these two disciplines. To simplify the process of identifying appropriate interdisciplinary connections, the *NGSS* include suggestions for linking the *CCSS* for both mathematics and English language arts to each science content topic.

The lesson that follows reflects three-dimensional dimensional science instruction by addressing disciplinary core ideas, crosscutting concepts, and science practices. Although it is targeted to a particular grade level, the lesson can be adjusted to address the progression of concepts for other grades with related disciplinary core ideas. The lesson includes suggestions for integrating engineering, mathematics, and language arts concepts.

It is not reasonable to expect one lesson to address all performance expectations or teach all the content related to a disciplinary core idea. The example here can be part of a series of lessons that build on one another to promote a deep understanding of the standard.

Earth and Space Model Lesson

NGSS Standard: 1-ESS1-1: Use observations of the Sun, Moon, and stars to describe patterns that can be predicted.

Disciplinary Core Idea: ESS1.A: Patterns of the motion of the Sun, Moon, and stars in the sky can be observed, described, and predicted.

Science and Engineering Practices:
- Asking Questions and Defining Problems
- Observing
- Analyzing and Interpreting Data

Crosscutting Concept: Patterns: Patterns in the natural world can be observed, used to describe phenomena, and used as evidence.

Nature of Science Idea: Scientific knowledge assumes an order and consistency in natural systems and is based on empirical evidence.

Teachers can introduce the lesson by asking students to list objects they have observed in the sky with their own eyes. Answers to this question will establish the students' prior knowledge and reveal any misconceptions that they have. Appropriate responses include stars, Moon, Sun, and clouds. If responses include items such as rainbows, birds, or planes, teachers should tell the students that they are only talking about objects that are always in the sky, all day and night and every day of the year. As a class, the students then determine which of the objects they named should be removed from consideration. Some students may believe that the Moon should be removed because it isn't in the sky during the day. If that's the case, teachers should assure the students that objects can be in the sky even if you can't see them. They should remind students that, like the Sun, the Moon remains in the sky at all times.

Eventually, the students' list will include the Moon, Sun, and stars. Teachers can ask if these objects in the sky and space always look the same. Students will often refer to the different shapes of the Moon and the different positions of the Sun in the morning, afternoon, and evening. This is when teachers should introduce the crosscutting concept of patterns by asking relevant questions: Do the changes in the sky happen in a pattern? Do they repeat themselves each day or as days go by? Can we predict what will happen the next day or night?

Students often talk about the Sun moving in the sky, "coming up" in the morning and "going down" at night. Teachers can explain that it is really the Earth that is moving while the Sun is staying still. "It takes one year for the Earth to travel around the Sun," they might add. "And guess what else? Not only does the Earth move around the Sun, it is also spinning around at the same time! It takes one whole day, 24 hours, for the Earth to make one complete spin. So when they see the Sun 'come up' in the morning or 'go down' at night, it is really the Earth that is moving, not the Sun."

The concept of the Earth moving in relation to the Sun and spinning on its axis at the same time can be hard for young children to understand, so teachers should present them with experiences that help them to do so. For example, students can role-play the movement of the Earth around the Sun by moving in a circle around a big yellow ball or flashlight representing the Sun. As they move around the model of the Sun, the students should spin or twirl like ballet dancers.

Another way for students to make observations and collect evidence of the Earth moving in a regular pattern relative to the Sun is to have them measure the length of their shadows outdoors at different times of day. Students can use strings to measure their shadows, then hang the strings along a wall and compare them. In this way, they begin to make graphical representations of length well before they are capable of constructing numerical graphs. The science lesson thus integrates the *CCSS* for mathematics of describing measurable attributes of objects and comparing their relative sizes.

Next, teachers should next ask the students: "Now that you have observed that the length of our shadows change during the day in a regular pattern, can we use this information to tell time? How can we figure it out? Can we do an investigation to answer our question?" It is important to remind the students that investigations always begin with a question, and that scientists always come up with a plan to try to answer their questions. Giving the students time to brainstorm possible ways to answer the question in groups and then share their ideas with the class is essential. Teachers may decide to have each group try out its plan or give all groups one plan to follow.

In our example, Let's suppose that the class decides to investigate Sundials as a possible means to tell time using the movements of the Earth. Teachers can distribute the image in Figure 11.1 and assign students to use washable markers and a straw or pencil to transform their hands into Sundials.

Teachers first demonstrate how to write the numbers on their own hands, then direct the students to write the numbers on their hands themselves. If students have difficulty doing this, they can pair up to help one another. Alternatively, if they keep science notebooks, they can trace an outline of their hands and label it as in Figure 11.1 or directly on the handout of the image.

The investigation should begin first thing in the morning on a Sunny day so that students can observe changes in the shadows as the hours go by. Once students have prepared their Sundials, they go outside and

Figure 11.1. Handy Sun Dial

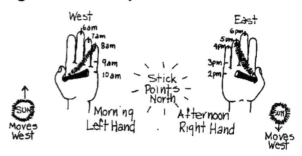

place a pencil between their thumb and the side of their left hand as shown in the diagram. Students should hold their palms out flat, with the pencil pointing up and toward the north. During the morning, students should tuck the pencil between their thumb and the side of their left hand and point their fingers to the west. In the afternoon, they should put the pencil in their right hand and point their fingers to the east. Children often need to be reminded to keep the pencil tightly tucked between their thumb and the side of their hands and to keep it pointing up and towards the north. Teachers should be prepared to face them in the correct direction. Once everything is in place, students will see a shadow form across their hands that extends to their fingertips. They should be instructed to observe which number on their hand the shadow touches. They can then make a drawing of the shadow as it appears on their hands on the handout or directly onto hand tracings in their science notebooks.

Teachers should take the students out at least five times during the day to collect data: twice in the morning, once at noon, and twice in the afternoon. The children should continue to observe the positions of the pencil's shadows and record their observations throughout the day. Each time the students repeat the activity, they should answer the following questions: *What did you observe? Was it the same or different from last time? What do you think you will see next time? Why do you think this?* Students should discuss possible reasons for the shadows changing position. If necessary, teachers should guide them to relate their observations to their measurements of the length of their own shadows. Finally, as a whole class, students should connect the activity to their earlier discussions about the movement of the Earth in relation to the Sun.

The next day, students repeat the activity, but this time they relate their observations to the time on a clock. Over the course of the day, the students should begin to observe patterns. One pattern is the way the shadow changes throughout the day; another is the correlation between these changes and the time on a clock. Teachers can use students' observations to discuss patterns in the Earth's movement relative to the Sun. The patterns that the students observe for themselves help them make sense of the motion of the Earth.

Students will need lots of additional reinforcing experiences to help them fully understand the Earth's movements in the sky. Ideally, this lesson should be repeated every other month to show the patterns of Earth's movement throughout the school year. Students can record their observations throughout the day and throughout the year. At the end of this activity, teachers can have students reflect on the patterns they observed as the shadow moved throughout the year and connect this phenomenon to the nature of science by explicitly noting that scientists also examine patterns to understand the world.

The concept of the Earth-Sun system is difficult for students to fully understand, especially first graders, and the *NGSS* recognizes the need to revisit the concepts again in later grades. This concept is housed in the overall Earth and Space Systems: Earth's Place in the Universe (ESS-1) standard. The activity discussed above can certainly be used again in grade 5 to address the following disciplinary core idea: "The orbits of Earth around the Sun and of the Moon around Earth, together with the rotation of Earth about an axis between its North and South Poles, cause observable patterns. These include day and night; changes in the length and direction of shadows; and different positions of the Sun, the Moon, and stars at different times of the day, month, and year."

Suggested Follow-up Instruction for the Education Methods Classroom

A preservice methods instructor could present the lesson above to students as an example incorporating the three dimensions of the *NGSS*: disciplinary core ideas, science and engineering practices, and crosscutting concepts. As such, the lesson can serve as a model for designing similar ones aligned to the K–6 *NGSS*. However, it can also be used to help preservice teachers

design and carry out more sophisticated science lessons touching on the nature of science, engineering connections, and connections to the common core.

Connecting to the Nature of Science

The fundamental premise behind the nature of science is that scientific knowledge assumes an order and consistency in natural systems and is based on empirical evidence. Research clearly shows that most preservice teachers do not understand NOS and consequently cannot teach it effectively (Lederman and Lederman 2014). A methods instructor could use the sample lesson as a starting point to introduce the nature of science and then ask the teaching candidates to brainstorm other ways to include NOS understandings in it and in their own lessons.

Connecting to Engineering Practices

To support engineering connections, consider that the construction of a sundial is essentially an engineering task. Other possible engineering problems for students to consider could include designing a more accurate sundial, one that you can wear, or a giant sundial in the schoolyard.

Connecting to the CCSS

Preservice teachers often need to draw meaningful connections for students between the science topics they are teaching and mathematics and English language arts concepts covered elsewhere in the curriculum. The sundial lesson contains several natural connections to mathematics and English language arts. For example, regarding mathematics, students are asked to measure, collect, and analyze data as well as to reflect on the concept of telling time. Specific *CCSS* for mathematics that connect to this lesson include 1.MD.A.1, "Order three objects by length; compare the lengths of two objects indirectly by using a third object," and 1.MD.B.3, "Tell and write time in hours and half hours using digital and analog clocks."

Preservice teachers could be asked to come up with other activities that meet both sets of standards, such as to investigate other devices for telling time, creating a timeline of the history of sundials, or learning to tell time by relating the sundial to an analog clock.

The lesson also connects to the English language arts *CCSS* W.1.7, "Participate in shared research and writing projects," and W.1.8, "With the guidance and support from adults recall information from experience or gathering information from a provided source to answer a question."

Working with trade books related to the science concepts that students are learning about is a popular classroom strategy, but simply reading about science ignores the three-dimensional focus of the *NGSS*. A methods instructor could use the sundial lesson to demonstrate effective use of trade books in three-dimensional science instruction, and as with the mathematics connection, preservice teachers could be asked to design language arts activities that augment inquiry-based science instruction. Some examples of young children's books about the Earth-Sun system, telling time, and shadows include *The Grouchy Ladybug* (1977) by Eric Carle, *How Do You Know What Time It Is?* (2013) by Robert E. Wells, and *The Sun's Day* (1989) by Mordicai Gerstein.

Having students use science notebooks to record observations, inferences, and questions is one way to connect lessons to writing standards. A methods instructor can extend the example to introduce other ways of using science journals, such as to list instructions for creating a sundial, write a report on the history of using the sky to tell time, or draw a timeline showing when different time-telling devices were invented.

References

American Association for the Advancement of Science. 1993. *Benchmarks for science literacy.* New York: Oxford University Press.

Bybee, R.W. 2013. *Translating the NGSS for classroom instruction.* Arlington, VA: NSTA Press.

Carle, E. 1977. *The grouchy ladybug.* New York: Harper Collins.

Gerstein, M. 1989. *The Sun's day.* New York: Harper Collins.

Lederman, N. G., and J. S. Lederman. 2014. Research on teaching and learning of nature of science. In *Handbook of research on science education.* Vol. 2, ed. N. G. Lederman and S. K. Abell, 600–620. New York: Routledge.

Lederman, N.G., and J.S. Lederman. 2016. Do the ends justify the means? Good question: But what happens when the means become the ends? *Journal of Science Teacher Education* 27 (2): 131–136.

National Research Council (NRC). 1996. *National Science Education Standards.* Washington, DC: National Academies Press.

NGSS Lead States. 2013. *Next Generation Science Standards: For states, by states.* Washington, DC: National Academies Press. *www.nextgenscience.org/next-generation-science-standards.*

Wells, R. E. 2013. *How do you know what time it is?* New York: Scholastic.

Norman G. Lederman *is distinguished professor of mathematics and science education at the Illinois Institute of Technology in Chicago and was an award-winning K–12 biology and chemistry teacher. Internationally known for his scholarship on students' and teachers' conceptions of the nature of science and scientific inquiry, he has authored and edited numerous books, chapters, and articles and made more than 500 national and international conference presentations. He was co-editor of both volumes of the* Handbook of Research on Science Education *and* School Science and Mathematics *and is currently co-editor of the* Journal of Science Teacher Education. *Lederman is a fellow of the American Association for the Advancement of Science and of the American Educational Research Association, and he received the Distinguished Contributions to Science Education through Research Award from the National Association for Research in Science Teaching. He can be reached by e-mail at* ledermann@iit.edu.

Judith Lederman *is associate professor and director of teacher education in the Department of Mathematics and Science Education at the Illinois Institute of Technology. She has been responsible for more than 600 national and international presentations and publications on the teaching and learning of scientific inquiry and the nature of science in both formal and informal settings. She is co-editor of the* Journal of Science Teacher Education *an author for a National Geographic elementary science textbook* series, and coauthor of an elementary science teaching-methods text. Lederman has served in South Africa as a Fulbright fellow and on the board of NSTA. She is also a past president of the Council for Elementary Science International. She has received such honors as Rhode Island State Teacher of the Year, the Milken Foundation National Educators Award, and a Christa McAuliffe Fellowship.

Selina L. Bartels is currently an assistant professor of science education at Concordia University Chicago and obtained her PhD in science education at the Illinois Institute of Technology. She has taught elementary, middle, and secondary school for 10 years in Chicago Public Schools. During her time as a teacher, she taught science, social studies, and reading. Bartels has written and delivered several publications and presentations focusing on young students' understandings about science and scientific inquiry.

CHAPTER 12

Constructing Explanatory Arguments Based on Evidence Gathered While Investigating Natural Phenomena in a Methods Course for Middle School Teachers

Frackson Mumba, Laura Ochs, Alexis Rutt, and Vivien M. Chabalengula

The effective implementation of the science and engineering practice "engaging in argument from evidence" will depend on teachers' experiences constructing explanatory arguments based on evidence, their understanding of the process of scientific argumentation, and their pedagogical knowledge about the role of argumentation in science teaching. In this chapter, we discuss ways to engage preservice science teachers in evidence-based argumentation and discuss how to assess their science content knowledge, argumentation skills, and ability to plan instructional activities centered on argumentation. Finally, we review lessons learned and further suggestions for engaging teachers in evidence-based argumentation.

Introduction

Science is social in nature, and advancements in scientific knowledge are most often achieved through collaboration among scientists (Asterhan and Schwarz 2007; McDonald 2010). Through collaboration, scientists use the process of argumentation to evaluate competing scientific ideas and to arrive at conclusions about natural phenomena (Ozdem et al. 2013). Engineers also engage in argumentation as they investigate natural phenomena, test design solutions, and use evidence to evaluate their solutions.

A Framework for K–12 Science Education (*Framework*; NRC 2012) and the *Next Generation Science Standards*

(*NGSS*; NGSS Leads States 2013) have identified construction of explanatory arguments based on evidence as one of the eight essential science and engineering practices. According to the *NGSS*, scientific argumentation is the process used to develop evidence-based conclusions and explanations. In addition, the practice supports critical thinking and promotes a deeper understanding of science content knowledge and the nature of science (Cavagnetto 2010). Because the collaborative and social nature of the argumentation process appeals to many students (Osborne 2010), it is an effective strategy for motivating them to learn about science.

Sampson and Blanchard (2012) found that teachers cited lack of time, low student ability levels, and their own lack of knowledge about the argumentation process as common obstacles to implementing the practice in their classes. Many teachers receive no formal training in scientific argumentation in their undergraduate science courses or preservice science methods courses. If training in evidence-based argumentation is present in teacher education at all, it usually focuses on *teachers'* argumentation skills (e.g., Aydeniz and Ozdilek 2015) or on argumentation as an instructional strategy (e.g., Simon, Erduran, and Osborne 2006). It is rare for teachers to receive firsthand instruction on the scientific use of explanatory argumentation (e.g., Asterhan and Schwarz 2007; Kaya 2013). This is unfortunate, because the willingness and ability of teachers to implement the practice in classrooms depends largely on their personal experience with and understanding of it.

Clearly, there is a need to engage preservice teachers in the construction of explanatory arguments based on evidence (e.g., McNeill and Knight 2013). There are four main rationales for doing this:

1. It helps teachers to develop an understanding of the need for empirical evidence in scientific inquiry.

2. It offers teachers insights about the process through which scientists develop new knowledge.

3. It shows teachers how to engage their students in constructing scientific explanations based on evidence.

4. It teaches preservice educators how to develop and utilize instructional materials for teaching the practice.

Description of Science Methods Courses

We engage our preservice teachers in constructing explanatory arguments in two consecutive science methods courses. The first course, in the fall semester, is designed to increase preservice teachers' understanding of science content knowledge, inductive instructional approaches, scientific argumentation, the nature of science, technology integration, and assessment. After preservice teachers have learned about the essential features of inquiry, they are ready to learn about argumentation.

In the inquiry lessons, teachers learn to use data analysis to formulate explanations. We present each instructional strategy by teaching a science lesson to preservice teachers, who participate as students, then debriefing the lesson by highlighting characteristics of the strategy. Next, the preservice teachers develop their own lesson plans or activities for each instructional strategy, then work in pairs to teach their lessons either in our class or during their clinical experiences at local schools.

The spring semester science methods course is designed to teach preservice teachers about project-based and problem-based learning, argumentation in engineering design solutions, and integration of engineering design into science teaching. The course also teaches how to incorporate these strategies into science units. This course addresses several of the science and engineering practices listed in the *NGSS* by using an informed engineering design approach (Burghardt and Hacker 2004) to engage preservice teachers in design projects that integrate science and engineering. Emphasizing the intelligent nature of engineering design helps motivate students to learn science and engineering concepts.

The activities in the course focus on specific STEM concepts. For example, a challenge might require students to design models for minimizing energy transfer as a way of learning about thermodynamics concepts. Once the context, specifications, and constraints of the problem have been made clear, the participants engage in short, focused activities related to the relevant content knowledge. Next, they use this knowledge to design a solution to the problem. Through the process of addressing the design challenge, participants also learn about the ways in which engineers use argumentation to investigate phenomena and how to test solutions using evidence to evaluate the claims made by others.

Instructional Models for Engaging Teachers in Scientific Explanatory Arguments

Scientists and engineers generate new knowledge in two main ways: by collecting, analyzing, and interpreting their own data, and by analyzing and interpreting data collected by others. We engage our preservice teachers in these two processes by using modified versions of the Generate an Argument and Evaluate Alternatives models developed by Sampson and Gerbino (2010).

The main difference between the two models is that the Generate an Argument model has learners use data collected by others, whereas the Evaluate Alternatives model has them collect the data themselves. Both models require participants to analyze the data to answer research questions. Because these models were developed for use with K–12 students, we have modified both of them to meet the needs of preservice teachers by adding three steps: an initial explanation, a connection to scientific knowledge, and an assessment of content knowledge, argumentation skills, and the ability to use the model to plan a lesson.

Modified Instructional Model for Generating an Argument

Step 1: Identify the problem and research question. Small groups of preservice teachers are given a handout that includes a description of the problem, scientifically oriented research questions to be answered, and instructions for presenting the evidence to support teachers' claims and explanations. The teachers are asked to identify the problem and research questions presented in the description.

Step 2: Generate a tentative claim. Each group uses prior knowledge to develop a tentative argument. These argument statements are presented on a poster that shows the tentative claim, evidence, and research question.

Step 3: Develop an initial explanation. Each group uses prior knowledge to provide explanations that support their tentative claims. While formulating their initial explanations, groups are not allowed to use any outside resources. The goal is to identify background and related knowledge as well as any misconceptions.

Step 4: Analyze the data and formulate an explanation. Each group analyzes the data provided by the instructor. Sometimes explicit analysis instructions are provided; other times, it is up to the group to determine which data are most important and what types of analysis are needed. Then, each group interprets its analysis and formulates explanations that support or refute the tentative claim.

Step 5: Connect the explanations to scientific knowledge. Each group uses additional scientific resources to compare accepted scientific knowledge to the explanations formulated in Step 4. Summaries of related theories or resources may be supplied by the

instructor or teachers may be tasked with locating them. The goal is for teachers to determine whether their claims and explanations are consistent with accepted scientific explanations.

Step 6: Engage in argumentation. Using a round-robin structure, groups present their claims, evidence, and explanations to the class. One person in each group stays with the poster and presents the group's argument while the others circulate to learn about the arguments developed by other groups (Sampson and Berdino 2010). Participants communicate their ideas and evidence, evaluate explanations, and ask or answer questions. This step helps preservice teachers to understand that the goal of scientific argumentation is not to win, but to develop a better understanding of the scientific concepts under investigation.

Step 7: Reflect. After the presentations, groups review the peer feedback and any new evidence they may have encountered to revise their claims and explanations. We wrap up the instruction by summarizing the problem or research question, the nature of the data analyzed, and the claims presented by the groups before explicitly connecting the activity to the main science concepts that the lesson was designed to address. Preservice teachers are then required to write individual summaries of the main science concepts employed throughout the activity and to describe how they might modify the activity for middle or high school students.

Step 8: Assess content knowledge and argumentation skills. The revised claims submitted by each group and the summaries submitted by individual participants are scored to evaluate argumentation skill, understanding of the science content, and the ability to use this model to plan lessons. (This step is discussed in further detail following the description of the second instructional model.)

For a lesson using this model to study climate change, a handout might read as follows:

> Stories abound in the media about the effects of climate change on our planet. Extreme weather, melting glaciers, and habitat loss are just a few of the consequences that scientists attribute to the continuous warming of our climate. Although most scientists agree that our climate is changing, reasons for this change remain a source of contention. Some scientists claim that the change in climate is part of a natural cycle of the Earth's warming and cooling. Other scientists argue that climate change as we see

CHAPTER 12

it now is largely caused and accelerated by human activity. Your team of scientists has been charged with analyzing the data provided to evaluate these two competing explanations of the cause of climate change.

The guiding research question for this lab is: "What causes climate change?" Your argument should include a claim about the reasons for climate change, evidence to support your claim, and a justification of your evidence. Due to the contentious nature of this question, it is imperative that you use the data provided to you. You also have access to the internet and the school library to locate any additional data that supports your claim.

Once you have made your claim, compiled your evidence, and written a justification of the evidence, you must organize the research question, your claim, your evidence, and the justification on a display board. Make sure that your claim is well supported and that the evidence you are using makes sense and is closely related to your claim. Be sure to compare your explanations to accepted scientific knowledge about climate change

We will use a round-robin approach to share the arguments and claims. One person in your group will remain at your table to share your argument while the remaining members will circle around and critique other arguments. As a critic, you will need to determine the credibility of other groups' arguments based on their evidence, explanation, sources, and justification. Does their explanation make sense? Do they provide enough supporting evidence to convince you that their explanations are empirically based and consistent with accepted scientific knowledge on climate change?

After completing the round robin, you will be provided with additional instructions for submitting a reflection of your individual experience with this activity.

Modified Instructional Model for Evaluating Alternatives

Step 1: Introduce the scientific phenomenon. The instructor describes observations pertaining to a phenomenon and provides several possible explanations for them.

Step 2: Write or review research questions. Participants are either asked to use the information presented in the first step to write research questions that will guide testing of the possible explanations, or they are provided with a set of research questions to review and possibly refine.

Step 3: Select a claim and provide an initial explanation of it. Groups are asked to choose one of the claims and use prior knowledge to provide supporting explanations for it. While selecting a claim to support, the groups are not allowed to use outside resources. The goal is to identify their background and related knowledge as well as any misconceptions.

Step 4: Gather evidence. The groups design experiments and collect data to test their claims. To ensure that all safety issues have been considered, each group's experiment is approved by the instructor.

Steps 5–9 of this model are the same as steps 4–8 of the modified Generate an Argument model.

For a lesson using this model to study carbon dioxide, a handout might read as follows:

Scientists have many concerns about the increasing amounts of greenhouse gases, like carbon dioxide, that are getting into our atmosphere. But what exactly does carbon dioxide do?

The guiding research question for this lab is "How does carbon dioxide affect the Earth's atmosphere?" Scientists have provided two competing answers to this question. One group of scientists claims that atmospheric carbon dioxide operates as a thermostat that controls the temperature of Earth. The other group attributes carbon dioxide gas to natural causes, such as global temperature increase. In your groups, evaluate these claims and select one to support with evidence. You will justify this explanation using supporting evidence to the class.

First, however, you will design and conduct an investigation for assessing the difference between a closed system with heavy amounts of carbon dioxide and a closed system with typical atmospheric air. Lab supplies include three dry 16-ounce water bottles, clay, a straw, baking soda, vinegar, and wireless temperature probes. You will also have access to sunlight.

Once you have designed and completed your investigation, use your data and any background knowledge you have to make a claim about the research question. Don't forget to provide convincing evidence and a justification. After making your claim, compiling your evidence, and writing a justification, you must organize the research question, your claim, your evidence, and the justification for your evidence on a display board. Make sure that your claim is well supported and that the evidence you are using makes sense and is closely related to your claim.

We will use a round-robin approach to share our arguments. One person in your group will remain at your table to share your argument while the remaining members circle around and critique others' arguments. As a critic, you need to determine the credibility of other groups' arguments based on their evidence and justification. Does it make sense? Do they provide enough supporting evidence to convince you that what they are saying is true?

Table 12.1 shows that these example labs not only engage preservice teachers in constructing explanatory arguments based on evidence but also address other science practices, core ideas, and crosscutting concepts.

Assessment

Our assessment focused on establishing the extent to which preservice science teachers had increased their understanding of the scientific phenomena under investigation, improved their argumentation skills, and gained pedagogical knowledge about the use of argumentation in science teaching. To assess understanding of science content, we use a model developed by Zohar and Nemet (2002) that involves examining the quality and accuracy of the content knowledge presented in the participants' claims, evidence, justifications, and explanations. The quality of explanations and justifications are scored as follows: *no consideration of scientific*

Table 12.1. Matrix of Argumentation Activities and the *Framework*

A Framework for K–12 Science Education	Climate Change	CO₂ Lab
Science and Engineering Practices		
Asking questions	X	X
Developing and using models		
Planning and carrying out investigations		X
Using mathematics and computational thinking	X	X
Constructing explanations	X	X
Engaging in argument from evidence		
Obtaining, evaluating, and communicating Information	X	X
Crosscutting Concepts		
Patterns	X	X
Cause and effect: Mechanism and explanation	X	X
Scale, proportion, and quantity		X
Systems and system models	X	X
Energy and matter: Flows, cycles, and conservation	X	X
Structure and function		
Stability and change	X	
Disciplinary Core Ideas		
Human impacts on Earth systems	X	X
Global climate change	X	X
Natural resources	X	X

knowledge, inaccurate scientific knowledge, nonspecific scientific knowledge, or *correct scientific knowledge.*

To assess preservice science teachers' argumentation skills, we use a model developed by Sampson and Gerbino (2010) that involves determining the extent to which the empirical evidence is relevant and fits with the claim, the extent to which the claim is sufficient and consistent with accepted scientific theories and laws, and whether the data analysis was conducted using appropriate methods.

To assess teachers' ability to plan instruction centered on argumentation, we require each of them to use the two instructional models described in this chapter to develop two science lesson plans of their own. We then analyze their plans and assess their fidelity to the models. We also use open-ended prompts to assess understanding of science content knowledge and scientific argumentation and of the difference between scientific argumentation and scientific explanation.

Conclusions, Challenges, and Recommendations

We have learned that engaging preservice science teachers in explanatory arguments based on evidence increases their science content knowledge, argumentation skills, understanding of the nature of science, and ability to plan instruction centered on argumentation. We have also observed that preservice teachers tend to demonstrate higher levels of interest and motivation while learning about instructional strategies centered on scientific argumentation and that some preservice teachers engage in more argumentation if the phenomena under investigation are applicable in K–12 science classrooms. However, some teaching candidates demonstrate resistance to learning science through argumentation, and others exhibit challenging knowledge gaps when it comes to science concepts. It is also difficult to develop activities that are relevant to all teachers regardless of grade level or science discipline. We try to overcome these challenges through individual accommodations and flexible instruction. Often, we are able to leverage the challenges to initiate discussions about interdisciplinary core ideas and prompt deeper reflection among prospective teachers.

References

Asterhan, C. C., and B. B. Schwarz. 2007. The effects of monological and dialogical argumentation on concept learning in evolutionary theory. *Journal of Educational Psychology* 99 (3): 626-639.

Aydeniz, M., and Z. Ozdilek. 2015. Assessing preservice science teachers' understanding of scientific argumentation: What do they know about argumentation after four years of college science? *Science Education International* 26 (2): 217–239.

Burghardt, M. D., and M. Hacker. 2004. Informed design: A contemporary approach to design pedagogy as the core process in technology. *Technology Teacher* 64 (1): 6.

Cavagnetto, A. R. 2010. Argument to foster scientific literacy: A review of argument interventions in K–12 science contexts. *Review of Educational Research* 80 (3): 336–371.

Kaya, E. 2013. Argumentation practices in the classroom: Preservice teachers' conceptual understanding of chemical equilibrium. *International Journal of Science Education* 35 (7): 1139–1158.

McDonald, C. V. 2010. The influence of explicit nature of science and argumentation instruction on preservice primary teachers' views of nature of science. *Journal of Research in Science Teaching* 47 (9): 1137–1164.

McNeill, K. L., and A. M. Knight. 2013. Teachers' pedagogical content knowledge of scientific argumentation: The impact of professional development on K–12 teachers. *Science Education* 97 (6): 936–972.

National Research Council (NRC). 2012. *A framework for K–12 science education: Practices, crosscutting concepts, and core ideas.* Washington, DC: National Academies Press.

NGSS Lead States. 2013. *Next Generation Science Standards: For states, by states.* Washington, DC: National Academies Press. *www.nextgenscience.org/ next-generation-science-standards.*

Osborne, J. 2010. Arguing to learn in science: The role of collaborative, critical discourse. *Science* 328 (5984): 463–466.

Ozdem, Y., H. Ertepinar, J. Cakiroglu, and S. Erduran. 2013. The nature of pre-service science teachers' argumentation in inquiry-oriented laboratory context. *International Journal of Science Education* 35 (15): 2559–2586.

Sampson, V., and M. R. Blanchard. 2012. Science teachers and scientific argumentation: Trends in views and practice. *Journal of Research in Science Teaching* 49 (9): 1122–1148.

Sampson, V., and F. Gerbino. 2010. Two instructional models that teachers can use to promote and support scientific argumentation in the biology classroom. *The American Biology Teacher* 72 (7): 427–443.

Simon, S., S. Erduran, and J.Osborne. 2006. Learning to teach argumentation: Research and development in the science classroom. *International Journal of Science Education* 28 (2/3): 235–260.

Frackson Mumba *and* **Vivien M. Chabalengula** *are associate professors of science education at the University of Virginia. Mumba's research on science teacher education and science teaching and learning has been published in several journals and book chapters, and he has made more than 170 presentations at science education conferences. Chabalengula's research on preservice science teachers and student learning has also resulted in several journal articles and book chapters.*

Laura Ochs *and* **Alexis Rutt** *are doctoral students at the University of Virginia. Ochs's research deals with engineering design in science instruction, and Rutt's research focuses on technology-integrated argumentation inquiry in science classrooms for English language learners. Contact the authors at* fm4v@eservices.virginia.edu.

CHAPTER 13

Building a Foundation for Three-Dimensional Instruction Through Bridging Practices in a Secondary Methods Classroom

Julie A. Luft and Robert Idsardi

In this chapter, we share methods for preparing preservice teachers to uphold the vision of the *Next Generation Science Standards* (*NGSS*; NGSS Lead States 2013). We begin by discussing bridging practices related to content, adapting lessons, data-driven decisions, and collaborating with colleagues. To illustrate how these practices can be used with preservice teachers, the authors also share examples from their own experience. Ultimately, our goal is not to advocate for specific practices but to promote the idea that those who work with preservice teachers should deliberately engage them in activities aligned with the *NGSS*.

Bringing the *NGSS* to the classroom will be a formidable task for our newest science teachers. While they are still learning to teach, they will have to modify the existing curriculum and develop activities and assessments that align with updated objectives. In addition, they will likely be put in the position of assisting more experienced teachers who are also modifying their instruction to meet the requirements of the *NGSS*. These new responsibilities will challenge even the most able newly hired science teachers. As future instructional designers and leaders, the next generation of science teachers will need coursework on science instruction, instructional materials, and student learning as well as practical training on implementing the *NGSS*.

Bridging practices can help early-career science teachers negotiate the difficult transition from pre-service programs through the first years of teaching. These practices guide teacher instructional decisions by focusing on lesson content, the learning of all students, and the data acquired from instruction. Furthermore, bridging practices promote the ideals of the *NGSS* among all teachers, whether new or experienced, by encouraging collaboration to improve lesson design and execution.

Background

In one of the few papers to discuss the early-career development of educators, Feiman-Nemser (2001) recognized the importance of central tasks in teaching. These tasks should connect directly to the classroom, she asserted, and they should help to strengthen and sustain instruction. More recent research suggest that Feiman-Nemser's model should more clearly emphasize subject matter knowledge as integral to central tasks (Luft 2009; Luft et al. 2011). In this way, early-career teachers can focus on understanding the content that they teach while also making sound decisions about instruction of that content (Luft 2012).

Using central tasks in the different phases of early-career teacher learning can have a positive effect on teacher professional development (e.g., Ball and Forzani 2009; Windschitl, Thompson, and Braaten 2012). Ball and Forzani (2009) refer to these tasks as *high-leverage practices.* They have developed a suite of 19 such tasks, including eliciting student ideas, establishing classroom discourse norms, and setting up and managing small-group work. All of these practices are cross-disciplinary and solidly grounded in education theory (TeachingWorks, 2013). However, it is not clear how teachers employing this model cultivate an understanding of the content or learn about practices that are tailored to their specific content areas.

Windschitl, Thompson, and Braaten (2012) refer to central tasks as *ambitious practices,* and they use them to support the teaching of model-based inquiry in science. They focus on four practices: crafting instruction around big ideas, eliciting student thoughts, making sense of activity, and pressing students for evidence-based explanations. Teachers have been shown to carry these practices forward into the classroom.

Bridging practices emphasize the three dimensions of learning: disciplinary core ideas, scientific and engineering practices (which includes models), and crosscutting concepts. An overview of the three dimensions can be found in Table 13.1.

Following are the bridging practices we emphasize for bridging the gap between preservice learning and the first years of classroom teaching while strengthening and sustaining instruction.

Content Considerations

The content knowledge of science teachers has a significant effect on the quality of instruction (e.g., Abell 2007; Van Driel, Berry, and Meirink 2014). Many researchers and teacher educators leave the content knowledge preparation of science teachers to science faculty. This assumes that teachers require the same content knowledge as scientists, which research suggests is not true (Deng 2007; Shulman 1986). In fact, content considerations have already led to a greater understanding of the unique knowledge needed to teach mathematics (Ball, Lubienski, and Mewborn 2001).

Table 13.1. The Three Dimensions of Learning in the *NGSS*

Dimension	Important Components	Examples
Disciplinary core ideas	• Have broad importance across many disciplines or are a key organizing concept • Provide a tool for understanding or investigating complex ideas • Have varying levels of depth and sophistication over time	• Ecosystems: interactions, energy, and dynamics • Motion and stablity: forces and interactions • Earth's place in the universe
Crosscutting concepts	Apply across all domains of science	• Patterns, similarity, and diversity • Scale, proportion, and quanitity • Systems and system models
Science and engineering practices	The practices that scientists and engineers use to investigate and explain the natural world or to design and build models and systems	• Asking questions • Developing and using models • Engaging in argument from evidence

As with Ball, Thames, and Phelps's (2008) domains of Mathematical Knowledge for Teaching (MKT), the practice of content considerations includes three components:

1. *Core content knowledge* (CCK) is the core knowledge that teachers are expected to have about the science topics they will teach as identified by the *NGSS*.

2. *Specialized content knowledge* (SCK) is the science knowledge teachers require to understand the difficulty that students may have with a content topic.

3. *Linked content knowledge* (LCK) is an understanding of the connections between science topics or concepts and the ways in which they relate to or build on each other.

In the methods classroom, preservice secondary science teachers are repeatedly prompted to consider CCK, SCK, and LCK as they plan their lessons. Figure 13.1 (p. 110) provides an outline of these prompts from Luft and Nixon (forthcoming). These types of prompts force preservice teachers to consider the content they will be teaching and methods of presenting that content that support student learning.

As preservice teachers begin to engage in more structured fieldwork, they have constant discussions about content focused on the CCK, SCK, and LCK of lessons. Explicitly considering lesson content is a practice that will persist as teachers begin teaching in the classroom.

Adapting Lessons

Although new teachers often want to create lessons from scratch (Luft et al. 2011), it is important for them to understand how to adapt lessons as well. Many educators use the 5E Instructional Model for this purpose (Bybee et al. 2006), which has the following five components: Engage, Explore, Explain, Elaborate, and Evaluate. Studies have shown this model to enhance student learning (Wilson, Taylor, Kowalski, and Carlson 2010).

The 5E Model is aligned with the scientific and engineering practices found in the *NGSS*, which include asking questions, developing models, carrying out investigations, analyzing and interpreting data, using mathematical thinking, constructing explanations, engaging in argument, and communicating information. The model is also well suited for integrating crosscutting concepts into science instruction. These are bridging ideas that span disciplines, including patterns; cause and effect; scale, proportion, and quantity; systems and system models; energy and matter; structure and function; and stability and change (NGSS Lead States 2013). During the methods course, instructors modify existing lessons to match the format of the 5E Model and to draw on the scientific practices and the crosscutting concepts outlined in the *NGSS*. Preservice teachers then use these modified lessons with middle school and high school students during their student teaching experiences. During the first years of teaching, teachers can draw upon their skills adapting lessons to better tailor their instruction to the learning of all students and the *NGSS*.

Making Data-Driven Decisions

Classroom instruction should be informed by an accurate assessment of student knowledge. Analyzing student learning data can help to guide instructional decisions and inform evaluation of the curriculum. Teachers should share the data with students so they can have insights into their learning. Using data to improve instruction is an essential classroom practice (Mandinach 2012).

Drawing on the work of Love (2008), we encourage preservice teachers to ask questions about their students' performance and then collect the data necessary to answer them. Examples of questions include the following: *How well did all my students understand the concepts at the end of the unit? What difficulties did students have in understanding the concepts? What instructional approaches seemed most effective for supporting all my students' learning?*

Teachers analyze the data they collect in preservice teacher groups to make decisions about the structure and focus of their next lessons. In some cases, they will need to review material or supplement their instruction. The data may also suggest supports teachers might consider for struggling students. Students who are just learning English may need additional assistance with the language of science, for example, or students with visual impairments may need specially designed materials. Once secondary science teachers embark on their

Figure 13.1. Content Consideration Prompts for Preservice Teachers

\<Discipline\> Concept Sketch

A concept sketch illustrates the main aspects of your knowledge of a concept or system. Add details to the diagram below to create a concept sketch that represents \<the concept\>. The details you add should show relationships among components and processes. Include labels, important theories, laws, additional models, or explanations as needed.

\<two macro components of the concept—one material and one energy source\>

Circle the area(s) in the top diagram that are generally difficult to understand. Explain your thinking in this box.

What knowledge/ concepts precedes learning about \<concept\>?	What knowledge/ concepts follows learning about \<concept\>?

If needed, use the back of this paper.

first years of teaching, data-driven decisions should be commonplace for them. Use of this practice can help newly hired teachers better understand the problems they encounter in classrooms and shift their mindset from one of managing a class to one of managing student learning.

Collaborating With Colleagues

When teachers work collaboratively, they reinforce, build, expand, and challenge their notions about teaching science. In their early review of professional development research, Wilson and Berne (1999) reported that successfully supporting the use of new practices among teachers required collaboration among peers and within educational communities. In a more recent analysis of science and mathematics teacher data from the United States Eisenhower Mathematics and Science Education program, Garet et al. (2001) found that collaboration within a school, grade level, or subject was an important feature of professional development programs. The practice of collaborating with colleagues supports purposeful conversations among science teachers about logistical, instructional, philosophical, and psychological needs (Luft, Bang, and Roehrig 2007).

Logistical discussions focus on identifying resources and procedures needed to teach science. These include discussions about school procedures, finding materials for laboratories, and planning field trips. Instructional discussions focus on the knowledge necessary for instruction, such as how to adapt or sequence lessons or how to analyze student work. Philosophical discussions help teachers build foundational beliefs and knowledge related to practice. Psychological discussions consist of emotional support and creating an identity within a community.

During preservice coursework, the emphasis is on cultivating instructional and philosophical discussion. These discussions continue during student teaching, when logistical and psychological discussions are introduced. By the time teachers are in their first years of independent teaching, they are familiar with all four kinds of discussion and prepared to collaborate with colleagues to support their learning and instruction.

Figure 13.2. A Model Representing the Combination of Bridging Practices and the *NGSS*

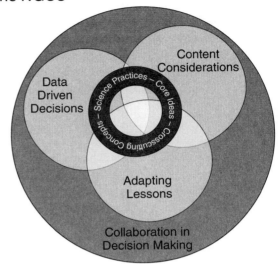

Examples From the Methods Classroom

Our secondary methods course is one of three courses that preservice teachers take prior to their student teaching, the other two being a course on curriculum and a field component. The field component is in a district with a high percentage of ethnic minorities, all of whom qualify for free meals. Because families from low socioeconomic backgrounds often lack the time and financial resources to support student learning at home, teaching can be especially challenging under these circumstances—making them a perfect testing ground for bridging practices.

We embed our Bridging Practices throughout the methods course, as in the following examples. (See Figure 13.2.)

Exploring the Progression of Understandings Through Grades K–12 per the NGSS

Early in the semester, preservice teachers are grouped according to area of expertise—life sciences, physical sciences, or Earth and space sciences—and asked to predict the level of understanding that students will need to have of a disciplinary core idea by grade 12. The teachers then analyze the progression of expected understanding of the disciplinary core idea for each grade from 1 to 12 per the *NGSS*. A week later, each student brings to class a document outlining this progression and shares whether their work confirmed their initial prediction about twelfth-grade understanding of the core idea. Many teachers are surprised by the level of understanding specified in the *NGSS*, which is often more conceptual and less detailed than the standards to which they're accustomed. The preservice teachers then use what they've learned in this exercise to complete the content considerations handout. Preservice teachers are required to think critically about the content that they know and the content that they will be teaching during this exercise as well as to engage in instructional and philosophical discussions.

Adapting Lessons for All Students

In this exercise, preservice science teachers learn to adapt lessons to include scientific practices aligned to the *NGSS*. Teachers usually engage in the activity toward the middle of the semester, after they have learned about the *NGSS* and the 5E Model.

We begin with a traditional science lesson in which teachers are given a question to answer along with related data and a data table. The lesson occurs in the classroom so that teachers can experience how teacher-oriented it is. All teachers work on the same part of the lesson at the same time.

Next, students are asked to discuss their learning experiences. Most report that the lesson is boring and not very challenging. They are also asked to discuss how the lesson may be perceived by English language learners (ELLs) who may have limited experience with language related to the topic. This discussion builds on our prior readings and discussions about diversity.

Students are then asked to work in their teams to adapt the lesson for two different groups: a class of middle-level ELLs who are just learning how to make scientific explanations and a class of secondary students who are new to the topic but experienced with related scientific practices. Teachers are required to give a rationale for all their modifications. After teams complete their modified lessons, teams take turns presenting them to the class and students engage in a whole-group critique of each one.

Final Comments

The *NGSS* will be ubiquitous in science classrooms in upcoming years. To implement them successfully, preservice science teachers need to understand the standards and how to enact them in their first years on the job. Learning about and engaging in bridging practices is one way to ensure that they do this.

We designed our bridging practices to support our preservice science teachers as they engage both in field experiences and in their first years of teaching. In developing the strategy, we considered the specific instructional needs of science teachers in our area and chose to accommodate some practices already in place. Many schools hold regular meetings to examine student performance using different data sources. Learning how to participate in these meetings should happen during preservice education rather than the first years of teaching.

Bridging practices reinforce the value of continuous teacher development over time. Preservice teachers will not have all the knowledge and skills they need in their first years; rather, they will build their knowledge and practices as they interact with students, colleagues, and the curriculum (Luft, Dubois, Nixon, and Campbell 2015). Using these bridging practices with teachers before they're hired can help them to learn how to build their knowledge from the start.

References

Abell, S. K. 2007. Research on science teacher knowledge. In *Handbook of research on science education*, ed. S. K. Abell and N. G. Lederman, 1105–1149. Mahwah, NJ: Lawrence Erlbaum Associates.

Ball, D. L., and F. M. Forzani. 2009. The work of teaching and the challenge for teacher education. *Journal of Teacher Education* 60 (5): 497–511.

Ball, D. L., S. T. Lubienski, and D. Mewborn. 2001. Research on teaching mathematics: The unsolved problem of teachers' mathematical knowledge. In *Handbook of research on teaching*. Vol. 4, 433–456. Washington, DC: American Educational Research Association.

Ball, D. L., M. H. Thames, and G. Phelps. 2008. Content knowledge for teaching: What makes it special? *Journal of Teacher Education* 49 (5): 389–407.

Bybee, R. W., J. A. Taylor, A. Gardner, P. Van Scotter, J. C. Powell, A. Westbrook, and N. Landes. 2006. *The BSCS 5E Instructional Model: Origins and effectiveness*. Colorado Springs, CO: BSCS.

Deng, Z. 2007. Transforming the subject matter: Examining the intellectual roots of pedagogical content knowledge. *Curriculum Inquiry* 37 (3): 79–295.

Feiman-Nemser, S. 2001. From preparation to practice: Designing a continuum to strengthen and sustain teaching. *Teachers College Record* 103 (6): 1013–1055.

Garet, M. S., A. C. Porter, L. Desimone, B. F. Birman, and S. K. Yoon. 2001. What makes professional development effective? Results from a national sample of teachers. *American Educational Research Journal* 38 (4): 915–945.

Love, N. 2008. *Using data to improve learning for all: A collaborative inquiry approach.* Thousand Oaks, CA: Corwin.

Luft, J. A. 2009. Beginning secondary science teachers in different induction programmes: The first year of teaching. *International Journal of Science Teaching* 31 (17): 2355–2384.

Luft, J. A. 2012. Subject-specific induction programs: Lessons from science. In *National Society for the Study of Education, 111th yearbook,* ed. T. Smith, L. Desimone, and A. Porter, 417–442. New York: Teachers College Press.

Luft, J. A., E. Bang, and G. H. Roehrig. 2007. Supporting beginning science teachers. *The Science Teacher* 74 (5): 24–29.

Luft, J. A., S. Dubois, R. Nixon, and B. Campbell. 2015. Supporting newly hired teachers of science: Attaining professional teaching standards. *Studies in Science Education* 51 (1): 1–48.

Luft, J. A., J. B. Firestone, S. S. Wong, I. Ortega, K. Adams, and E. Bang. Beginning secondary science teacher induction: A two-year mixed methods study. *Journal of Research in Science Teaching* 48 (10): 1199–1224.

Luft, J. A., and R. Nixon. Forthcoming. Subject matter knowledge that is needed to teach science: Conceptualizing and exploring this construct. *International Journal of Science Education.*

Mandinach, E. B. 2012. A perfect time for data use: Using data-driven decision making to inform pratice. *Educational Psychologist* 47 (2): 71–85.

NGSS Lead States. 2013. *Next Generation Science Standards: For states, by states.* Washington, DC: National Academies Press. *www.nextgenscience.org/ next-generation-science-standards.*

Shulman, L. S. 1986. Those who understand: Knowledge growth in teaching. *Educational Researcher* 15 (2): 4–14.

TeachingWorks. 2013. High-leverage content. *www. teachingworks.org/work-of-teaching/high-leverage-content.*

van Driel, J. H., A. Berry, and J. Meirink. 2014. Research on science teacher knowledge. In *Handbook of research on science education,* ed. N. G. Lederman and S. K. Abell, 848–870. New York: Routledge.

Wilson, C. D., J. A. Taylor, S. M. Kowalski, and J. Carlson. 2010. The relative effects and equity of inquiry-based and commonplace science teaching on students' knowledge, reasoning, and argumentation. *Journal of Research in Science Teaching* 47 (3): 276–301.

Wilson, S. M., and J. Berne. 1999. Teacher learning and the acquisition of professional knowledge: An examination of research on contemporary professional development. *Review of Research in Education* 24: 173–209.

Windschitl, M., J. Thompson, and M. Braaten. 2012. Proposing a core set of instructional practices and tools for teachers of science. *Science Education* 96 (5): 878–903.

Julie A. Luft *is the athletic association professor of science and mathematics education in the College of Education at the University of Georgia. As a science teacher–education researcher, she focuses on secondary science teachers and faculty who teach undergraduate courses. She has published numerous research articles and chapters, and she has edited five books. Over the years, she has served the field of science education as the research director of NSTA, an associate editor of many journals, and president of ASTE. As an academic, she has received various research and mentoring awards, the most recent of which was a Fulbright Award to work in Vietnam. She can be contacted via e-mail at* jaluft@uga.edu.

Robert Idsardi *is a doctoral student in science education at the University of Georgia. His research focuses on ways to support science educators at the K–12 and undergraduate levels in their use of active learning techniques. More specifically, he is interested in exploring how teachers' views of their students' learning influence classroom instruction.*

Undergraduate Science Courses for Preservice Science Teachers

CHAPTER 14

Engaging Prospective Teachers in Learning Disciplinary Core Ideas and Crosscutting Concepts in an Undergraduate Course in Biological Science or Biochemistry

Ann T. S. Taylor and Joseph W. Shane

In this chapter, we discuss ideas for embedding aspects of the *Next Generation Science Standards* within biology courses that prospective teachers commonly complete at the undergraduate level. We begin by reviewing the guidelines presented in *Vision and Change in Undergraduate Biology Education* (*Vision and Change*; AAAS 2009) for developing and executing preservice biology classes, then discuss ways of aligning *Vision and Change* guidelines to the *Next Generation Science Standards* (*NGSS*; NGSS Lead States 2013). We also review the ways in which the *NGSS* align with the *Standards for Science Teacher Preparation* (*SSTP*; NSTA 2012).

Introduction

In 2006 and 2007, the American Association for the Advancement of Science (AAAS), with support from the National Science Foundation (NSF), convened an advisory board of 16 distinguished biology professors and college administrators to lead a working group of NSF personnel, pedagogy experts, researchers, faculty, students, and representative from the field to develop a shared vision for college-level biology education and to identify the changes needed to achieve that vision. This work culminated in the *Vision and Change* guidelines. Like the *NGSS*, *Vision and Change* calls for integrating core concepts across the curriculum and focusing on student-centered learning.

Comparing *Vision and Change* and the *NGSS*

Both *Vision and Change* and the *NGSS* provide lists of core concepts that every student should learn during their biology classwork. As shown in Table 14.1 (p. 118), the guidelines are very similar. The major difference is that *Vision and Change* includes an additional category, Pathways and Transformations of Energy and Matter,

Table 14.1. Comparison of *Vision and Change* and *NGSS* Core Concepts

Vision and Change	*NGSS*
Structure and Function: Basic units of structure define the function of all living things.	**LS1:** From Molecules to Organisms: Structures and Processes
Systems: Living systems are interconnected and interacting.	**LS2:** Ecosystems: Interactions, Energy and Dynamics
Pathways and Transformations of Energy and Matter: Biological systems grow and change by processes based on chemical transformation pathways and are governed by the laws of thermodynamics.	Divided between **LS1** and **LS2**
Evolution: The diversity of life evolved over time by processes of mutation, selection, and genetic change.	**LS4:** Biological Evolution: Unity and Diversity

which centers on metabolic signaling and developmental pathways as well as the underlying chemical, thermodynamic, and genetic bases of these processes.

The *Vision and Change* model does not have a recommendation that parallels the *NGSS*'s crosscutting concepts. Though standards for K–12 science education have historically included students' understanding of interdisciplinary concepts, these are rarely points of emphasis in post-secondary science programs. However, *Vision and Change* does acknowledge the idea of overarching and interconnecting themes with the category Tap into the Interdisciplinary Nature of Science, which connects disparate areas within biology and the biological sciences to other scientific fields.

Another key distinction, as shown in Figure 14.1, is that *Vision and Change* statements are broader and more general than those in the *NGSS*. Although both documents emphasize that it is the role of students to ask questions and define problems, plan and carry out investigations, analyze and interpret data, construct explanations and design solutions, and engage in argument from evidence, *Vision and Change* aggregates all these skills into one broad category, whereas the *NGSS* articulates each one more precisely. What's more, the *NGSS* further divides the communication competency to address professional and general audiences (i.e., "Communicate and collaborate with other disciplines" and "Understand the relationship between science and society").

The above differences notwithstanding, both *Vision and Change* and the *NGSS* promote roughly the same instructional methodology. Both documents emphasize

acquiring knowledge through active learning strategies and the use of scientific processes, and both stress applying these techniques early and often. Collaborating with science teacher education faculty can help instructors of preservice teachers to implement these reforms. Such collaboration has the added benefit of improving communication between campus departments.

The *NGSS* and the *Standards for Science Teacher Preparation (SSTP)*

To teach content effectively, teachers must be able to do more than merely understand science concepts. They must also be able to design and implement lessons and assessments that are appropriate for their students' developmental levels. It is for this reason that we recommend undergraduate institutions consider education coursework and licensure requirements.

Although not all programs are required to implement them, we use the NSTA's (2012) *Standards for Science Teacher Preparation (SSTP)* to raise awareness of professional expectations for training future science teachers. The *SSTP* are often used as part of an institution's accreditation under the guidelines of the Council for the Accreditation of Educator Preparation. Like the *NGSS*, the *SSTP* have an extensive research base and were designed in collaboration with organizations such as the American Association of Physics Teachers, the National Earth Science Teachers Association, the American Chemical Society, and the National Association of Biology Teachers.

Figure 14.1. Comparison of *Vision and Change* Disciplinary Practices (circles) and *NGSS* Science and Engineering Practices (rectangles)

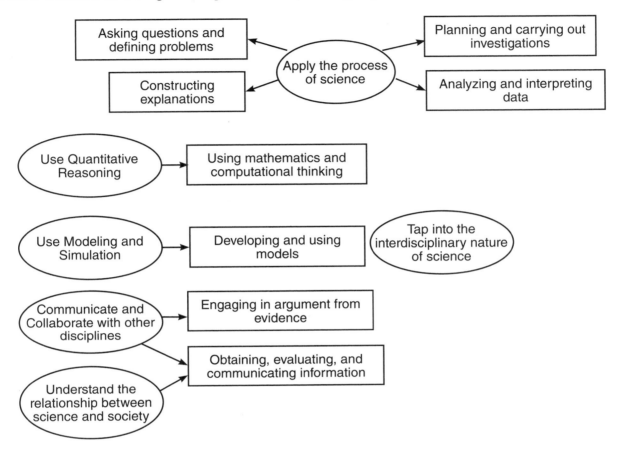

The *SSTP* are organized around the following six broad standards: (1) Content Knowledge, (2) Content Pedagogy, (3) Learning Environments, (4) Safety, (5) Impact on Student Learning, and (6) Professional Knowledge and Skills.

Standard 1: Content Knowledge

This standard refers to (a) preservice science teachers' understanding of the major ideas within their primary and related disciplines, (b) their use of science-specific instructional technology, and (c) their application of state- and national-level standards within their areas of licensure. In each scientific discipline, the NSTA requires that high school teachers demonstrate specific core, advanced, and supporting content competencies. In biology, for example, the core competencies are likely to be quite familiar: taxonomy, ecology and population dynamics, genetics, cell structure and function, environmental health, evolution, basic biochemical pathways, and the use of theoretical models. Advanced competencies are more granular and include pathology, social and historical issues, advanced genetics and heredity, and the ability to conduct research. Supporting competencies include biochemistry, chemistry laboratory skills, and selected topics in physics and Earth science. Separate sets of more basic competencies have been developed for middle school science teachers and elementary science specialists.

The NSTA has developed a content analysis form that teacher preparation programs use to align the required coursework at their institutions with the core, advanced, and supporting competencies. These forms, along with major-specific GPA data and passing rates on standardized content exams such as the Praxis II examinations, are common sets of data that programs provide in order to demonstrate that they attain Standard 1. (As of this writing, the NSTA is in the process of revising the content analysis form to align with the *NGSS*.)

Standard 2: Content Pedagogy

This standard requires preservice science teachers to use research-based learning theories to transform their personal content knowledge into a variety of age-appropriate, inquiry-based lessons. These lessons should take students' preconceptions into account and use science-specific technology (e.g., instrumentation, computer-interfaced probes) to support student learning. Following implementation of the *NGSS*, we expect that this standard will be updated to include science and engineering practices.

Standard 3: Learning Environments

This standard is perhaps the broadest of them all. It requires teachers to design lessons that are fair, equitable, and safe for all students and to use a variety of assessments that complement different teaching strategies. Science teacher preparation programs are required to use a comprehensive unit plan to provide evidence that they are addressing Standards 2 and 3. In biological science, preservice teachers typically design unit plans around a traditional theme such as ecology, genetics, or cell structure and function. This is the most comprehensive assessment used for accreditation purposes, and it is intended to illustrate a wide range of teaching strategies that can be adapted to any science classroom.

Standard 4: Safety

This standard focuses on chemical safety, emergency equipment and procedures, and the legal and ethical treatment of living organisms. Preservice teachers are required to have a personal understanding of each aspect of safety and to write lessons that explicitly teach and model safe practices for students. Comprehensive safety training is often not included in undergraduate science curricula, and this standard provides an excellent opportunity for preservice teachers to work with colleagues in science teacher education. Safety procedures must be included in preservice teachers' unit plans as well as in the instruments used to evaluate science student teachers.

Standard 5: Impact on Student Learning

This standard extends the assessment aspect of Standard 3 by requiring preservice science teachers to provide evidence that they have gauged students' understanding and scientific skills throughout instruction. They must also demonstrate that they are using the assessment data to modify and improve subsequent instruction. Broad assessment portfolios are often used during student teaching or other practicum experiences to demonstrate that this standard is being met.

Standard 6: Professional Knowledge and Skills

This standard refers to preservice teachers' involvement in both scientific and education professional development activities such as conferences, research, and community projects. Teacher preparation programs typically require students to provide examples of their involvement along with narrative explanations that describe the ways in which these experiences are consistent with this standard.

There are, naturally, numerous areas of direct overlap between the *SSTP* and the *NGSS*. The first dimension of the *NGSS*, science and engineering practices, aligns well with Standards 1 and 6, particularly with regard to the use of models and experience with genuine scientific research. Additionally, the requirement that preservice science teachers design inquiry-based instruction is consistent with the instructional techniques promoted by the *NGSS*.

The disciplinary core ideas for life science in both the *NGSS* and *Vision and Change* are consistent with the core, advanced, and supporting competencies of the *SSTP*. There is, however, less overlap with crosscutting concepts, as the revised *SSTP* tends to focus more on discipline-specific content knowledge.

Comparing the expectations of frameworks such as the *NGSS*, *Vision and Change*, and *SSTP* is valuable for assessing the content and pedagogical strategies used in postsecondary courses and programs. However, it is important to bear in mind that these frameworks represent guidelines, not prescriptions.

Recommendations for Undergraduate Biological Science and Biochemistry Courses

The following recommendations can be implemented in biological science and biochemistry courses to align them with Vision and Change and the NGSS.

Course-Specific Recommendations

Backward design

This strategy entails planning course activities by starting with the learning goals first. Although most college instructors have not had any educational training, scientific societies and funding agencies broadly support initiatives to increase the use of effective learning strategies in collegiate classrooms (Holm, Carter, and Woodin 2001). Models such as Scientific Teaching (Handlesman, Miller, and Pfund 2007) and Learning Cycles (Withers 2016) emphasize backward design and active learning techniques.

Case-based instruction

In this strategy, students must apply course concepts to a real-world scenario to develop an explanation or solve a problem. Cases often require students to apply skills such as analysis of laboratory and clinical data, construction of arguments, use of models, and communication of information. These skills correlate with the *NGSS* practices and *Vision and Change* competencies. Cases can also address ethics (Fisher and Levinger 2008), historical context (Allchin 2000), societal issues (Chamany 2006), and common student misconceptions (Cliff 2006). The National Center for Case Study Teaching in Science curates a collection of over 600 peer-reviewed cases that can be searched by subject, education level, topic, or type (NCCSTS 2017). At the extreme, entire courses can be taught thematically, using broad topics such as health, environment, energy, or food (Labov, Reid, and Yamamoto 2010). This coursewide approach emphasizes integration across core concepts and disciplines, such as mathematics and other areas of scientific study.

A particularly interesting adaptation combines case-based instruction and systems-based model construction. Speth and Bray (2006) provide an example of a lesson for a large, introductory biology course that uses model-based instruction to actively engage students in learning about the process through which genes determine phenotypes. This systems-based approach uses peer review and revision of models to help students in a large classroom setting construct their own explanations and share understanding. Additionally, this type of learning activity can grow in complexity as the students advance. Although lower-level learning objectives can be met through simple or didactic models, accomplishing more complex or in-depth learning objectives often requires students to build their own models from physiological cases.

Emphasis on literature

This strategy can be incorporated into coursework either sporadically in the form of individual exercises that focus on a single concept (Cornely 1999) or continuously throughout the entire theme or course. The latter approach is known as CREATE, which stands for *consider, read, elucidate, hypothesize, analyze and interpret data,* and *think of the next experiment* (Hoskins 2007). A CREATE module consists of four sequential articles from the same laboratory, which allows students to observe the progression of science. CREATE sequences for courses ranging from high school to upper level biology majors are available in an online repository (see *www.teachcreate.org*), though some follow the original model more closely than others (Hoskins 2017).

Near-peer mentors

This strategy is particularly effective at institutions with a large population of prospective teachers. Near-peer mentors are students who have recently completed a course and mentor students who are currently taking it. In peer-led team learning, near-peer mentors facilitate cooperative learning groups or supplemental instruction sessions (Presler 2009). The system reinforces learning for the peer mentors by forcing them to review and reconsider the course material while providing current students with an additional source of explanations. Often, students understand peer-mentor explanations better than those of instructors, perhaps because the former are more aware of misconceptions and concepts that are particularly elusive or easily overlooked. The strategy also allows for the use of more active pedagogies, especially in large classes (Allen 2005). This can help students gain real-world experience dealing with group dynamics and providing feedback.

Course-based undergraduate research experiences (CURE)

The ultimate integration of content and practices happens in the research laboratory. However, the personnel and equipment cost associated with these authentic investigations make it difficult to provide, especially for students not considering a profession in the sciences. One solution is the course-based undergraduate research experiences (CURE) strategy, which provides students with opportunities to conduct authentic research during the laboratory portion of a course (Bell et al. 2016). Although some courses lend themselves more easily to the use of lessons based on inquiry and scientific practices than others, most postsecondary science programs include several that can be adapted to include these methods.

Many additional sources, such as NSTA journals, offer additional strategies for modifying coursework to align with *NGSS* practices. It is vital that these strategies be explicitly modeled during science teacher preparation programs so that teachers can eventually faithfully reproduce them in the classroom.

Interdepartmental and Institutional Recommendations

It is beneficial for science and education faculty to meet periodically and review expectations for content and pedagogy together. Here are some related strategies.

Comparing approved curriculum sheets for preservice science teachers with the content expectations described in the *NGSS* and *SSTP*

This is a straightforward and worthwhile activity. At one of our institutions, we recognized that there were no requirements for future biology teachers to complete the upper-division course on evolutionary theory or environmental science. We quickly added these requirements. Given the interdisciplinary focus of the *NGSS*'s crosscutting concepts, this type of strategy will be increasingly important, and decisions are best made with the cooperation of all science departments.

Peer support for teaching and in-class field experiences

Peer-led tutoring and supplemental instruction related to work during practicum hours is economical and mutually beneficial, especially for institutions that already have a learning center in place providing undergraduate assistance.

Comprehensive safety training

In our experience, this strategy is all-too-often overlooked in both science and science education programs. The NSTA requires preservice science teachers to understand and implement broad aspects of science safety, but it is doubtful that any individual faculty member will have the expertise to properly oversee chemical storage and handling, use of safety equipment, and handling of animals and biohazardous materials. We recommend including comprehensive safety modules that include laboratory coursework, research, and other opportunities, such as seminars or freshman experiences. Another mutually beneficial format for addressing safety topics involves inviting local science teachers to participate in or assist with safety training at postsecondary institutions. These types of sessions provide ongoing professional development for practicing teachers and real-world perspective for prospective teachers, and they help maintain ties between postsecondary institutions and local schools systems.

Cocurricular professional development experiences

This strategy is becoming increasingly common. In our institutions, for example, we have expanded the traditional research seminars to include sessions about preparing for the job search, transitioning from undergraduate education to graduate school, and the differences between industrial research and academic research. We are now considering including pedagogy seminars as well. If possible, we recommend establishing an NSTA Student Affiliate chapter on campus and encouraging students to collaborate with student clubs in science departments. Additionally, post-secondary institutions should consider promoting campus-based corollaries to professional society initiatives, such as the biennial Transforming Undergraduate Education in the Molecular Life Sciences meeting (see *www.asbmb. org/SpecialSymposia/2017/education*) or the Biennial Conference on Chemical Education (see *www.divched. org/committee/biennial-conference-chemical-education*). Attending state and regional NSTA meetings is another valuable option for keeping both practicing and future educators up-to-date on current educational theories and techniques.

Experiences With Implementation

Although many education programs have been incorporating guidelines from external accreditation organizations for decades, external recommendations are fairly new to most biology professors. Consequently, although *NGSS*-aligned instructional techniques have been integrated into teacher preparation programs relatively quickly, they've taken longer to take hold in biology science courses. Many professors are simply not used to aligning their instruction to standards or incorporating science and engineering practices into their teaching. Complicating the process further, textbooks are not typically organized around these themes. Fortunately, faculty and others are already responding to these challenges. Organizations like the Partnership for Undergraduate Life Sciences Education (PULSE) have provided resources, recognition, networking, and training for faculty to support the implementation of *Vision and Change* goals, and textbooks that are organized around the core concepts and practices are beginning to make their way to the marketplace (Barsoum 2013).

These national trends are reflected at our own institutions and within our own fields. For example, Dr. Shane teaches primarily in secondary science education, and he has seen the *NGSS* incorporated alongside other standards. Dr. Taylor's institution has begun incorporating these standards and practices into individual courses, but programwide implementation has been delayed due to retirements and staff turnover. Waiting has allowed new faculty to participate in the discussion from the outset.

One common barrier to implementation is student response. At Dr. Taylor's institution, many biology majors who are interested in health-related careers take a junior-level biochemistry course that uses many of the practices and strategies recommended by the *NGSS* and *Vision and Change*. These students frequently remark that they would prefer lectures, although they also note that they do appreciate the case studies.

We believe that as the *NGSS* and *Vision and Change* are more broadly implemented, students will become more comfortable with the cognitive dissonance associated with active learning strategies and ultimately come to appreciate the greater depth of understanding that they offer.

Conclusion

Both *Vision and Change* and the *NGSS* help prepare scientists and future teachers for successful careers in their respective fields. These national guidelines allow colleges and universities to more comprehensively review their programs and focus on increasing awareness and support and modifying practices and expectations. Faculty need to be made aware of the two sets of standards and provided with support for aligning their lessons to them (e.g., from textbooks, national associations, or federal funding agencies). Faculty also need to explicitly connect class activities to the learning objectives for both content and skills so that prospective teachers can do the same for their future students.

References

Allchin, D. 2000. How not to teach historical cases in science. *Journal of College Science Teaching* 30 (1): 33–37.

American Association for the Advancement of Science (AAAS). 2009. *Vision and change in undergraduate biology education: A view for the 21st century.* Washington, DC: AAAS. *www.visionandchange.org.*

Bell, J., T. Eckdahl, D. Hecht, P. Killon, J. Latzer, T Mans, J. Provost, J. Rakus, E. Siebrase, and E. Bell. 2016. CUREs in biochemistry—Where we are and where we should go. *Biochemistry and Molecular Biology Education* 45 (1): 7–12.

Chamany, K. 2006. Science and social justice: Making the case for case studies. *Journal of College Science Teaching* 36 (2): 54–59.

Cliff, W. H. 2006. Case study analysis and the remediation of misconceptions abour respiratory physiology. *Advances in Physiology Education* 30 (4): 215–223.

Cornely, K. 1999. *Cases in biochemistry.* New York: John Wiley & Sons.

Fisher, E. R., and N. E. Levinger. 2008. A directed framework for integrating ethics into chemistry curricula and programs using real and fictioal case studies. *Journal of Chemical Education* 85 (6): 796–801.

Handlesman, J., S. Miller, and C. Pfund. 2007. *Scientific teaching.* Englewood, CO: Roberts and Company.

Holm, B., V. C. Carter, and T. Woodin. 2001. Vision and change in biology undergraduate education: Vision and change from the funding front. *Biochemistry and Molecular Biology Education* 39 (2): 87–90.

Hoskins, J. A. 2017. Welcome to CREATE. CREATE: Transform understanding of science. *www.teachcreate.org*.

Hoskins, S. G., L. M. Stevens, and R. H. Nehm. 2007. Selective use of the primary literature transforms the classroom into a virtual laboratory. *Genetics* 176 (3): 1381–1389.

Labov, J. B., A. H. Reid, and K. R. Yamamoto. 2010. Integrated biology and undergraduate science education: A new biology education for the 21st century? *CBE–Life Sciences Education* 9 (1): 10–16.

NGSS Lead States. 2013. *Next Generation Science Standards: For states, by states.* Washington, DC: National Academies Press. *www.nextgenscience.org/next-generation-science-standards*.

Presler, R. W. 2009. Replacing lecture with peer-led workshops improves student learning. *CBE–Life Sciences Education* 8 (3): 182–192.

Speth, A. R., and E. Bray. 2016. Beyond the central dogma: Model-base learning of how genes determine phenotypes. *CBE–Life Sciences Education* 15 (1): 1–13.

Tanner, D. A., and K. Tanner. 2005. Infusing active learning into the large enrollment biology class: Seven strategies, from the simple to the complex. *Cellular Biology Education* 4 (4): 262–268.

Withers, M. 2016. The college science learning cycle: An instructional model for reformed teaching. *CBE–Life Sciences Education* 15 (12): 1–12.

Ann T. S. Taylor *is chair of the Division of the Natural Sciences and Mathematics Departments at Wabash College. She regularly teaches biochemistry to upper-level science majors as well as nonscience majors. She serves on the Editorial Board for Biochemistry and Molecular Biology Education, and she regularly contributes to the National Center for Case Study Teaching in Science.*

Joseph W. Shane *is an associate professor of chemistry and science education at Shippensburg University. In addition to teaching chemistry and serving as department chairperson, he teaches undergraduate and graduate courses in science education, and he is codirector of a STEM Master of Arts in Teaching (MAT) program. He has also been involved with accreditation efforts at the national level through the National Council for Accreditation of Teacher Education and NSTA.*

CHAPTER 15

Engaging Prospective Teachers in Learning Disciplinary Core Ideas and Crosscutting Concepts in an Undergraduate Course in Chemistry

Sarah B. Boesdorfer

In this chapter, we discuss how firsthand experience learning chemistry content while engaged in science and engineering practices or crosscutting concepts helps preservice teachers experience the type of instruction they will eventually be expected to provide. We offer examples of such lessons in postsecondary chemistry instruction.

During the very first day of their collegiate-level general chemistry lecture course, prospective teachers and their classmates are shown the following series of demonstrations:

1. Opening a bottle of vinegar and observing that the odor quickly spread

2. Adding a drop of food coloring to water and watching it mix without stirring

3. Making white crystals (salt) "disappear" by adding them to water and stirring, and

4. Pouring together 50 ml of one liquid and 50 ml of another liquid, then measuring the resulting volume at 98 ml

The instructor claims to be "doing magic" when she makes things disappear or mix without stirring.

Then, she asks her class to describe a model of matter that would explain their observations. After discussing their ideas in small groups and sharing them out, the students establish a model for matter that states some variation of the following: (1) All matter is made up of tiny particles that we can't see with our eyes, (2) The particles are moving, and (3) There is empty space between the particles. Throughout the semester, students will add details to this model and use it to explain and understand behaviors of matter.

In this rather short activity, the instructor successfully introduced chemistry content by engaging her class in the science and engineering practice of developing and using models from *A Framework for K-12 Science Education* (*Framework*; NRC 2012) and *Next Generation Science Standards* (*NGSS*; NGSS Lead States 2013). The instructor consciously chose to use an *NGSS* practice as part of a teaching technique and require her class to take an active role in the learning process. These decisions help prospective teachers to both learn the content and to observe how the *NGSS* can be used to teach it.

This activity is one of many small steps I am taking to meet the *NGSS* in my teaching. We cannot expect prospective chemistry teachers to use the *NGSS* in their classrooms if the standards have never been modeled for them. The *NGSS* have only recently been adopted in some states, and other states have yet to officially adopt them (NSTA 2016). Even once adopted, the *NGSS* are implemented slowly, as past standards-based reforms have been (Southerland, Sowell, Blanchard, and Granger; Woodbury and Gess-Newsome 2002). Further, most chemistry teachers in the United States do not have chemistry, chemistry-teaching, or chemistry-related degrees (e.g., biochemistry or chemical engineering), and biology degrees are the most common type of nonchemistry degree held by high school chemistry teachers (Rushton et al. 2014; Smith 2013). This suggests that, in order to provide firsthand experience with SEPs and *NGSS* to the greatest number of prospective chemistry teachers, postsecondary science instructors need to identify the most effective opportunities within their teacher preparation programs and incorporate the use of these techniques into those courses.

Chemistry-Teacher Education

When candidates enroll in programs to become chemistry teachers, the pathway most often includes attaining a BA or BS in chemistry along with completing additional classes and exams required by their state for teaching certification. However, teachers typically enroll to pursue different specializations, such as biology or general science, and then receive an additional endorsement or a certification to teach chemistry. Most states require teachers to have taken three or four college-level chemistry courses to teach chemistry. There are some variations, however; for example, the state of Illinois requires teachers to complete 12 hours of chemistry and pass a chemistry-content test (ISBE 2016). Due to the credit hours required for certification or endorsement, it is reasonable to assume that all prospective chemistry teachers will take at least a general chemistry course sequence and perhaps a semester of organic chemistry. As it happens, a national survey of chemistry teachers shows that 96% of current chemistry teachers have

taken introductory chemistry and 83% have taken organic chemistry (Smith 2013).

No matter which path they take, prospective chemistry teachers will also complete methods courses in which they learn about effective teaching practices and how to employ them; 84% of current chemistry teachers have taken courses in science education (Smith 2013). In some of these courses, prospective teachers study teaching and learning across different disciplines; in others, they focus on content-specific methodologies. Exposing preservice teachers to *NGSS*-aligned practices makes it more likely that they will use the practices themselves in their future classrooms (Lotter, Harwood, and Bonner 2006; Loughran 2014).

Aligning to the *NGSS* in Methods Courses

In the two methods courses I teach to prospective chemistry teachers, I use activities that model how to teach content by having students engage in *NGSS*-defined practices. I often choose activities that also help preservice teachers clarify or solidify chemistry knowledge, not only because they will be teaching it soon, but also because they will need to pass a content test for teacher certification. These activities can just as easily be applied in a content course as in a method course. Two science and engineering practices that I emphasize in both my methods courses are constructing scientific explanations and engaging in argument from evidence. In one activity, I give prospective teachers instructions they will need to perform the "Drinking Candle" experiment (see instructions from the Planet Science website here: *www.planet-science.com/categories/experiments/magic-tricks/2012/05/drinking-candle.aspx*).

After they have observed, I ask them to write some questions about the experiment. Following a short discussion, each student picks a question and designs and performs an experiment to solve it. The structure of this discussion is adapted from Steps to Inquiry (Youth Science Canada 2011). Finally, students present the results of their experiments with their evidence to the class and provide a scientific explanation for the phenomena described. I have found when I use this activity in my method courses, students tend to struggle with offering clear scientific explanations that include a claim and supporting evidence while using scientific principles or concepts (McNeill and Krajcik

2011). The activity thus helps me determine students' abilities to use their chemistry knowledge to explain firsthand observations while also furthering their depth of knowledge.

I also like to have my students perform a slightly modified version of the experiment Reaction Rates: Why do Changes in Temperature and Reaction Concentration Affect the Rate of Reaction? (see Sampson et al. 2014). Many general chemistry courses use argument-driven inquiry to engage students in *NGSS*-aligned processes as they learn core content, which is "designed to provide students with an opportunity to develop their own methods to generate data, to carry out investigations, use data to answer research questions, write, and be more reflective as they work. Argument-driven inquiry, however, also integrates opportunities for students to engage in scientific argumentation and peer review" (Walker, Sampson, and Zimmerman 2011, p. 1049).

When my students perform the Reaction Rates experiment, they are using the *NGSS* practices to plan experiments, analyze data, and engage in arguments based on the evidence. As with the Candle in the Jar activity, this experiment helps to clarify science knowledge (here, the properties of dilution) and conceptual understanding (of rates of reaction at the particle level).

Unlike previous standards documents, *NGSS* explicitly includes engineering-related content and skills, so I believe it is important to also teach prospective teachers how to apply these to chemistry concepts. In this lesson, after a short activity to introduce the design process (see Lessoncast n.d.), students are asked to optimize the design of a can-crushing process (see Boesdorfer 2014), develop a reaction-powered car (see Rochefort, Momsen, and Hower 2004), or develop a process to plate copper onto another metal using only given materials (see TryEngineering 2016). Completing this activity requires preservice teachers to apply engineering principles and chemistry content knowledge, such as the assessment of gas laws or oxidation-reduction reactions and the use of stoichiometric calculations. Their understanding of the chemistry content therefore improves their understanding of teaching techniques that incorporate *NGSS* practices.

The activities described above are simply examples for prospective teachers, but given the vast array of skills and knowledge they must learn, they may also be the only experiences they have learning content while using SEPs. In their own classrooms, teachers frequently fall back on instructional techniques that mimic their own experiences as a student (Loughran 2014; Sarason 1990), so it is important for content and methods courses both to model these *NGSS*-aligned strategies.

Aligning to the *NGSS* in Content Courses

Including *NGSS* concepts and language in general chemistry courses helps guarantee that all prospective teachers experience chemistry through the lens of the new standards.

Using Models

One way I ensure this is by asking my general chemistry students to continually revisit the *NGSS*-aligned model of matter they develop in class and refine it by connecting it to new learning throughout the semester. For example, when we study chemical reactions, we relate them to our understanding of matter as being made up of particles. As students write and balance reactions, they do this as well.

Analyzing and Interpreting Data

Students must repeatedly use the *NGSS* practice of analyzing and interpreting data to develop an understanding of chemistry. For example, when teaching atomic theory, I present my students with photoelectron spectroscopy data and ask them to consider what it might tell us about the structure of the atom (see Bergman et al. 2013 for a complete description of this lesson). I do something similar when teaching periodic properties. I provide students with graphs of the data for atomic radius, ionization energy, and so on, then ask them to analyze the data in terms of patterns and trends that correlate to the arrangement of the periodic table.

I present students with data to analyze and interpret regularly, not only because it is an *NGSS* practice but also because I subscribe to constructivist theory, which holds that students must build their own understandings of concepts (Bodner 1986). Presenting students with data to analyze is one way to help them do this. Another effective constructivist strategy for engaging with this practice is Process-Oriented Guided Inquiry, or POGIL, which uses guided questions to help students analyze and interpret data to clarify

Table 15.1. The Science Writing Heuristic (SWH) as Adapted and Used in My Undergraduate Teaching Practice.

Instructor's Actions	Students' Actions
1. Provide students with a laboratory situation or problem.	None
2. Facilitate a class discussion to help students as the class creates their questions and procedures.	1. Ask *beginning questions* about the situation, and research any necessary *background information* about the experiment.
	2. Develop *procedures and tests* to answer the beginning questions.
3. Monitor and facilitate experimental procedures and data collection.	3. Perform the tests to collect *data and observations,* analyzing as appropriate with *calculations and graphs.*
4. Facilitate a class discussion of their claims and evidence.	4. Make a *claim* based on their *evidence and analysis* to answer their questions.
None	5. *Reflect* on their learning and answer a few *post-laboratory questions.*
5. Provide feedback on written reports.	6. Write a lab report that uses the italicized words above as the section headers for their report.

Source: Adapted from Del Carlo 2012 and Greenbowe and Hand 2005.

understanding of concepts (Farrell, Moog, and Spencer 1999; Moog et al. 2009).

Lab Experiences

Because I am not teaching a laboratory course, it is hard for students to plan and carry out investigations. To keep learning active despite this limitation, I have students design experiments using online simulations (e.g., the simulation on gas properties by the University of Colorado Boulder 2016). Although I provide students with questions rather than having them develop their own, they must manipulate (i.e., experiment with) the simulation to observe and collect data and then use that data to support their scientific explanations.

At a previous institution, when I taught smaller general chemistry courses that included a laboratory portion, I used the Science Writing Heuristic (SWH) for guided inquiry (Greenbowe, Rudd, and Hand 2005; Schroeder and Greenbowe 2008). In this technique, students are presented with data regarding a given situation and required to develop questions, design an experiment to answer those questions, present a claim supported by evidence, and explain their reasoning. Table 15.1 provides a basic overview of what I did and what my students did each week in the laboratory.

Although the SWH was developed prior to the *NGSS*, it emphasizes many of the same practices, such as asking questions, planning and carrying out investigations, analyzing and interpreting data, and constructing scientific explanations (Greenbowe et al. 2007; Poock, Burke, Greenbowe, and Hand 2007). It also requires students to learn and apply disciplinary core content knowledge. The technique is similar to argument-driven inquiry, and both strategies are currently being effectively employed in general chemistry laboratory courses (Greenbowe et al. 2007; Walker, Sampson, and Zimmerman 2011).

Another technique that engages students in SEPs while delivering or re-enforcing content is problem-based learning (PBL). The problem-based learning (PBL) approach presents students with a real-world problem that they must solve by using knowledge and skills that they develop in the process (Duch, Groh, and Allen 2001). Mataka and Kowalske (2015) researched the use of PBL in general chemistry labs where students are given information about a problem, such as detecting pesticides on produce or production of biodiesel, and required to formulate a question that would help them to identify a solution by designing and conducting lab experiments. Research indicates that PBL, which by its nature requires the use of *NGSS* practices, improves students' abilities to employ these practices and increases their understanding of core concepts (Tosun and Taskesenligil 2013).

Challenges of Incorporating *NGSS* Practices in General Chemistry Classes

Beneficial though it may be, incorporating the use of *NGSS* practices in undergraduate general chemistry classes also comes with numerous challenges. This is especially true of large, lecture-only courses, such as my current course, which consists of approximately 200 students, is taught in a traditional lecture hall, and has no laboratory component. This format makes it difficult to engage students in some of the practices, such as planning and carrying out an experiment or engaging in argument from evidence. Although I have found some effective solutions (using online simulations, for example), I have not yet found ways for my students to engage in all eight *NGSS* practices.

Additional challenges arise from the prerequisite nature of general chemistry courses, which requires specific content that cannot be altered to align with the practices, especially as they are not exclusively for prospective teachers. Fortunately, when paired with the appropriate content, the *NGSS* and related practices can benefit all students regardless of their career aspirations. Techniques that incorporate *NGSS* practices with content learning, like SWH, POGIL, and argument-driven inquiry, have been shown to teach chemistry concepts to all students as well as or better than traditional methods (e.g.,Conway 2014; Poock et al. 2007; Walker et al. 2012).

Final Thoughts

In my personal experience as both a methods and general-chemistry instructor, I am convinced that it is essential to provide prospective teachers with opportunities to use *NGSS* practices while learning chemistry content—a conviction well supported by research (e.g., Loughran 2014). Research on other reform-based movements and inquiry-based teaching methods show that experience as a student has a strong effect on the practices teachers use in their classrooms (Lotter, Harwood, and Bonner 2006; Loughran 2014).

Including these experiences in my courses has helped prospective teachers to learn chemistry content outside of a content course and increased the chances that they will use *NGSS* practices with their future students, whether they plan to teach chemistry exclusively or not.

Through incremental changes, it is possible for instructors at all levels to improve their curriculum and better align their instruction with the *NGSS*. For example, at the time of this writing, I am working on including *NGSS* practices and crosscutting concepts from the standards in verbal and written communications with my students. As Cooper (2016) argues, we need to say what we mean and speak the language of the *NGSS* in our courses. When I use a model to help students learn a chemistry concept and predict outcomes, I specifically refer to it as a model and ask students to use it to make predictions and to communicate ideas to others (see NGSS Lead States 2013, Appendix F). Although SWH, POGIL, argument-driven inquiry, and many other strategies that I recommend in this chapter were developed prior to the *NGSS*, they can be effective for implementing the standards, especially if teachers use *NGSS*-aligned language as they use the strategies.

My efforts to incorporate *NGSS* practices in my courses remain a work in progress. Ideally, there would be a chemistry content course for prospective science teachers of any discipline that would teach the content in context and emphasize the ones most relevant to specific careers, but this is unrealistic. In the meantime, I encourage postsecondary instructors to strive to incorporate SEPs into all chemistry courses to enhance the learning of chemistry content for all types of students while providing teaching majors with experience using effective *NGSS*-aligned instructional strategies.

References

Bergman, J. M. 2013. Teaching atomic theory using photoelectron spectroscopy data. *The Chemical Educator* 13: 1–5.

Bodner, G. M. 1986. Constructivism: A theory of knowledge. *Journal of Chemical Education* 63 (10): 873–878.

Boesdorfer, S. 2014. Can crush. *https://docs. google.com/document/d/19opodJJRM7VaV-neUC9_KCCaDMPGxOWgRVGEueE_LGo/ edit?usp=sharing.*

Conway, C. J. 2014. Effects of guided inquiry versus lecture instruction on final grade distribution in a one-semester organic and biochemistry course. *Journal of Chemical Education* 91 (4): 480–483.

Cooper, M. M. 2016. It is time to say what we mean. *Journal of Chemical Education* 93 (5): 799–800.

Del Carlo, D. August 2012. Personal communication.

Duch, B.J., S. E. Groh, and D. E. Allen. 2001. *The Power of problem-based learning: A practical "how to" for teaching undergraduate courses in any discipline.* Sterling, VA: Stylus Publishing.

Farrell, J. J ., R. S. Moog, and J. N. Spencer. 1999. A guided inquiry chemistry course. *Journal of Chemical Education* 76: 570–574.

Greenbowe, T. J., and B. Hand. 2005. Introduction to the science writing heuristic. In *Chemists' guide to effective teaching,* ed. N. J. Pienta, M. M. Cooper, and T. J. Greenbowe, p. 140. Upper Saddle River, NJ: Pearson Prentice-Hall.

Greenbowe, T. J., J. A. Rudd II, and B. M. Hand. 2007. Using the Science Writing Heuristic to improve students' understanding of general equilibrium. *Journal of Chemical Education* 84 (12): 2007–2011.

Illinois State Board of Education (ISBE). 2016. Illinois licensure, endorsement, and approval requirements. *www.isbe.net/licensure/requirements/endsmt_struct.pdf.*

Lessoncast. (n.d.) Marshmallow Challenge: Intro to engineering design process. *www.lessoncast.com/lesson/marshmallow-challenge-intro-to-engineering-design-process.*

Lotter, C. W., S. Harwood, and J. J. Bonner. 2006. Overcoming a learning bottleneck: Inquiry professional development for secondary science teachers. *Journal of Science Teacher Education* 17 (3): 185–216.

Loughran, J. J. 2014. Developing understandings of practice: Science teachers learning. In *Handbook of research on science education.* Vol. 2, ed. N .G. Lederman and S. K. Abell, 811. New York: Routledge.

Mataka, L. M., and M. G. Kowalske. 2015. The influence of PBL on students' self-efficacy beliefs in chemistry. *Chemistry Education Research and Practice* 16 (4): 929–938.

McNeill, K. L., and J. Krajcik. 2011. *Supporting grade 5–8 students in constructing explanations in science: The claim, evidence and reasoning framework for talk and writing.* New York: Pearson Allyn and Bacon.

Moog, R.S., F. J. Creegan, D. M., Hanson, J. N. Spencer, A. Straumanis, D. M. Bunce, and T. Wolfskill. 2009. POGIL: Process-Oriented Guided-Inquiry Learning. In *Chemists' guide to effective teaching.* Vol. 2, ed. N. J. Pienta, M. M. Cooper, and T. J. Greenbowe, 90. Upper Saddle River, NJ: Prentice Hall.

National Research Council (NRC). 2012. *A framework for K–12 science education: Practices, crosscutting concepts, and core ideas.* Washington, DC: National Academies Press.

National Science Teachers Association (NSTA). 2016. About the *Next Generation Science Standards.* NSTA. *www.ngss.nsta.org/About.aspx.*

NGSS Lead States. 2013. *Next Generation Science Standards: For states, by states.* Washington, DC: National Academies Press. *www.nextgenscience.org/next-generation-science-standards.*

Poock, J. R., K. A. Burke, T. J. Greenbowe, and B. M. Hand. 2007. Using the science writing heuristic in the general chemistry laboratory to improve students' academic performance. *Journal of Chemical Education* 84 8: (2007): 1371–1379.

Rochefort, S., E. Momsen, and J. Hower. 2004. Kitchen chemistry: The chemical reaction powered car. In *Dr. Skip's corner: K–12 teaching adventures @ OSU Engineering.* Corvalis, OR: Oregon State University. *www.engineering.oregonstate.edu/momentum/k12/march04/index.html .*

Rushton, G. T., H. E. Ray, B. A. Criswell, S. J. Polizzi, C. J. Bearss, N. Levelsmier, H. Chhita, and M. Kirchhoff. 2014. Stemming the diffusion of responsibility: A longitudinal case study of America's chemistry teachers. *Educational Researcher* 43 8: 390–403.

Sampson, V., P. Carafano, P. Enderle, S. Fannin, J. Grooms, S. A. Southerland, C. Stallworth, and K. Williams. 2014. *Argument-driven inquiry in chemistry: Lab investigations for grades 9–12.* Arlington, VA: NSTA.

Sarason, S. B. 1990. *The predictable failure of educational reform: Can we change course before it's too late?* San Francisco: Jossey-Bass.

Schroeder, J. D., and T. J. Greenbowe. 2008. Implementing POGIL in the lecture and the Science Writing Heuristic in the laboratory: Student perceptions and performance in undergraduate organic chemistry. *Chemistry Education Research and Practice* 9: 149–156.

Smith, P. S. 2013. 2012 National survey of science and mathematics education: Status of high school chemistry. Chapel Hill, NC: Horizon Research.

Southerland, S. A., S. Sowell, M. Blanchard, and E. M. Granger. 2011. Exploring the construct of pedagogical discontentment: A tool to understand science teachers' openness to reform. *Research in Science Education* 41 (3): 299–317.

Tosun, C., and Y. Taskesenligil. 2013. The effect of problem-based learning on undergraduate students' learning about solutions and their physical properties and scientific processing skills. *Chemistry Education Research and Practice* 14 (1): 36–50.

TryEngineering. 2016. Can you copperplate? TryEngineering. *www.tryengineering.org/lesson-plans/can-you-copperplate.*

Walker, J. P., V. Sampson, J. Grooms, B. Anderson, and C. O. Zimmerman. 2012. Argument-driven inquiry in undergraduate chemistry labs: The impact on students' conceptual understanding, argument skills, and attitudes toward science. *Journal of College Science Teaching* 41 (4): 74–81.

Walker, J. P., V. Sampson, and C. O. Zimmerman. 2011. Argument-driven inquiry: An introduction to a new instructional model for use in undergraduate chemistry labs. *Journal of Chemical Education* 88 (8): 1048–1056.

Woodbury, S., and J. Gess-Newsome. 2002. Overcoming the paradox of change with difference: A model of changes in the arena of fundamental school reform. *Educational Policy* 16 (5): 763–782.

Youth Science Canada. 2011. Smarter science steps to inquiry: Investigation design and perform. *www.smarterscience.youthscience.ca/sites/default/files/documents/smarterscience/2-1_CDN_EN_Initiate_and_Plan_L2.pdf.*

Sarah B. Boesdorfer is an assistant professor in the Department of Chemistry at Illinois State University. Her teaching experiences include general chemistry, methods courses for prospective science teachers, and graduate courses and professional development programs for inservice science teachers. Chemistry teacher learning and teaching practices drive her research interests, and her current focus is on learning experiences aligned with the NGSS and the ways in which these experiences can translate into common classroom practices. She also has experience as a high school chemistry and physics teacher in Illinois. She can be reached via e-mail at sbboesd@ilstu.edu.

CHAPTER 16

Constructing Arguments Based on Evidence Gathered While Investigating Natural Phenomena in an Undergraduate Course in Earth Science

Michael A. Gibson

In this chapter, we discuss an adaptable instruction activity used in undergraduate Earth science courses with preservice teachers to illustrate the gains in conceptual understanding that result from firsthand data collection and analysis. This activity connects sea-level change throughout history to local areas in which sea level cannot be directly observed and allows learners to construct a conceptual model of the Earth science idea of "deep time," while fostering opportunities for inquiry related to issues that have arisen over long spans of geologic activity. The type of interactive, phenomenon-based learning offered by this activity aligns with the *Next Generation Science Standards* (*NGSS*; NGSS Lead States 2013), and it can help teachers address controversial topics by allowing learners to arrive at their own conclusions.

Introduction

As well as being an academic topic, global climate change has become a hotly contested political issue (e.g., Breslyn, McGinnis, McDonald, and Hestness 2016; Hestness 2014; Holthuis 2014; Robertson and Barbosa 2015). Several polls, such as one conducted by the National Association of Geoscience Teachers (NAGT), have found that teachers are generally ill prepared to tackle this topic and are often reluctant to teach it in depth due to its controversial nature

(e.g., Evans 2016; Plutzer et al. 2016). Although the scientific community is largely in agreement over the reality of global change, it is common for members of the general population to fall victim to misconceptions that continue to feed the controversy (e.g., Robertson and Barbosa 2015). The public needs to have a basic understanding of scientific processes and principles if it is to address these misconceptions. The *NGSS* were developed with this goal in mind.

One of the primary objectives of the *NGSS* is to foster skills for constructing explanatory scientific arguments based on evidence gathered by students while investigating natural phenomena. Students are expected to gather data, analyze it scientifically, and reach testable conclusions about actual natural phenomena. Preservice education programs often include methods courses intended to train teachers to achieve this goal. In these courses, teaching candidates are encouraged to provide opportunities for students to manipulate materials and methods under controlled conditions. However, global change is a very large and complicated set of processes that interact both spatially and temporally, and these interactions can be difficult to model in a laboratory.

Understanding global climate change requires learners to engage the associated issues on many levels. Learners must first have a conceptual understanding of the basic science of change, including the temporal and spatial scales, and must be able to give evidence for change on these scales. It is only after mastering these basic principles that learners can explore how global climate change affects our society and the world at large (e.g., Trendell Nation, Feldman, and Wang 2015).

When students gather evidence, especially if it is collected from locations near their own homes, that information is connected to their daily lives. This instructional strategy is called the phenomenon approach, and it adds relevance to the lesson by allowing students to more easily understand how their lives are affected by the concepts being investigated (e.g., NGSS Lead States 2013; Smith, Barber, Duguay, and Whitley 2012; Trendell Nation, Feldman, and Wang 2015). Unfortunately, many people erroneously believe sea-level change to only affect the coasts, and that therefore only coastal schools can provide students with opportunities to directly observe the phenomenon (e.g., Baumgartner et al. 2008; Strang et al. 2010). Due to this misconception, most students do not get to investigate the effects of sea- level change through analysis of real data that they collect. Another challenge is that global change processes take too long to occur for the average person to witness completely. The effects of past global change are buried within the Earth's geologic record as rock and sediment, and more current effects are masked by cyclical trends and other processes.

Too often, students have trouble understanding the relevance of processes that they view solely as a part of the Earth's distant past. To make these issues more meaningful, it is helpful to frame global change in the context of a current, ongoing phenomenon. This strategy aligns well with the *NGSS*, which emphasize the investigation of natural phenomena as an instructional strategy. Natural phenomena are defined as observable events to which students can apply their knowledge to explain or predict effects, and investigation of these phenomena provides students with knowledge-building opportunities that facilitate the development of evidence-based, scientific hypotheses (NGSS Lead States 2016).

In these types of investigations, students do not just model scientific inquiry, but rather directly engage with actual phenomena. Earth science phenomena are particularly suited to this approach, because they occur on all spatial and temporal scales and do not have to be modeled or scaled down to be used in a laboratory. In the Earth sciences, phenomena include a host of processes, such as mineral growth and rock formation, physical and chemical weathering, erosion and deposition of sediments, earthquakes, volcanic eruptions, mass wasting, flooding, weather and storm events, organic evolution and extinction, Earth-Moon-Sun interactions, asteroid impacts, and all issues related to global change. Earth system approaches to the geosciences emphasize crosscutting concepts that occur across many of these processes.

To give preservice teachers experience with phenomenon-based instruction, I use an adaptable activity in my undergraduate Earth science course that combines interactive data collection and analysis of phenomena related to sea-level change. This activity connects changes in sea level throughout history to local areas where the process cannot be directly observed and engages learners in constructing a conceptual model of the Earth science idea of "deep time."

The investigation of phenomena related to global sea-level change can be used to illustrate many of the Earth systems disciplinary core ideas addressed in middle school curricula. These ideas include, but are not limited to, the history of Earth (ESS1) and large-scale system interactions (ESS2). These core ideas involve such crosscutting concepts as pattern recognition, cause and effect, system models, and stability versus change. The crosscutting concept of stability versus change is particularly easy to relate to the disciplinary core idea of Earth and human activity (ESS3), as the use of the sediment and fossil evidence in this activity directly engages students with investigations of climate change and, by extension, sea-level change. Furthermore, discussion of sea-level change connects to the core ideas of natural hazards (ESS3.B) and human impacts on Earth systems (ESS3.C). By focusing on the results of sea-level change, these investigations allow students to analyze and interpret sediment and fossil data to forecast future sea level and predict (catastrophic) events. Students then generate questions that help them make sense of evidence showing the results of rising temperatures and rising sea level over the past century (MS-ESS3-5) and earlier.

How the *NGSS* Address Global Climate Change

Global change issues are applicable to nearly all *NGSS* at nearly all grade levels. Through the standards, students explore the global scale of the causal mechanisms of change and their local effect on biogeography, biodiversity, physical characteristics of habitats through environmental parameters, material resources available for organisms, and geologic processes. Earth systems connections have always formed a common vertical thread in education standards, and global change is a phenomenon that cuts across all disciplines, not just the sciences. History, political science, literature, art, and many other content areas all relate to societal responses to environments and change (e.g., Munsart 1997).

The *NGSS* address global change and sea-level rise both directly and indirectly. For example, the *NGSS* for grade 2 describe a unit on Earth systems processes that shape the Earth, which states that students "who demonstrate understanding can use information from several sources to provide evidence that Earth events can occur quickly or slowly" (2-ESS1-1). Similarly, the middle school *NGSS* about natural selection and adaptations states that "students who demonstrate understanding can analyze and interpret data for patterns in the fossil record that document the existence, diversity, extinction, and change of life forms throughout the history of life on Earth under the assumption that natural laws operate today as in the past" (MS-LS4-1). As indicated by these standards, the fossil record of global change, including sea-level change, is data that can be analyzed to clearly establish the existence of these processes, as well as to provide data about the rate and direction of change. The middle school history of life standard (MS-ESS2-2) states that learners are "demonstrating understanding if students construct an explanation based on evidence for how geoscience

processes have changed Earth's surface at varying time and spatial scales."

As any geologist will assert, the record of change is archived in the rocks, sediments, and fossils preserved where they are found. Geological material appears everywhere, so phenomenological data is always present; all learners needed is to know how to gather it and how to read it. In accordance with *NGSS* goals, global change and sea-level change processes yield data that place modern short-term processes into historical and temporal frameworks. However, just as many students do not immediately realize that "deep time" geological processes continue to have an effect on the modern world, many teachers do not know how to make these connections apparent. Fortunately, preservice teachers of Earth science are usually expected to take a course in historical geology, where global sea-level changes over the past 700 million years serve as a central organizing theme. A quick internet search of "sea-level curves" will show results for no fewer than six global sea-level rises and drops of more than 200 meters over the past 600 million years and many hundreds of smaller sea-level changes (e.g., Levin 2013; Prothero and Dott 2010).

Exploring Evidence of Sea-Level Change in Landlocked States

Of course, gleaning information from a readily available chart is not as engaging as observing evidence of phenomena in the real world and holding it in your own hands. Thankfully, evidence of sea-level change is preserved all around us, even within the landlocked states of the central United States, in the form of ancient marine sediments and their preserved marine fossils. In the plains of Kansas, limestone preserves fossilized clams, oysters, cephalopods, and fish. The mere existence of these types of deposits is *prima facie* evidence of sea-level change at a grand scale—an even larger scale than the one-meter rise in sea level currently being projected by climate scientists.

In my undergraduate Earth science course, preservice teachers complete the following activity that requires them to gather data about sea-level change that can also be related to modern phenomena. The activity uses actual geologic materials from Tennessee and Virginia, but it is easily adapted to any location with fossiliferous marine strata, regardless of (current) distance from the coast. Although the activity has been used primarily with preservice middle school teachers, it is highly adaptable. Because it centers on using data to construct student-driven questions that culminate in explanatory arguments, the activity can be adapted to any grade level or science discipline. Although I won't get into much detail here for the sake of focus, a good summary of the science of sea-level change can be found in Coe (2003), and Davis (2011) is a good reference for societal issues directly related to sea-level change in the Gulf of Mexico.

Activity for Understanding Evidence of Sea-Level Change

Marine geologists and paleontologists (scientists who study past oceans) and marine scientists (oceanographers and marine biologists who study modern oceans) use a variety of methods to sample sediments and organisms from oceanic settings. Most collecting of sediment or organisms is done remotely from aboard a ship, but these collections may also be augmented with video from remotely operated vehicles or satellites. Sediment and benthic organism sampling devices that are used to remotely sample the sea floor include corers, grabs, and dredges. Corers (Figure 16.1a, b) penetrate the sea floor vertically and retrieve a tube-shaped core of the shallow sea-floor sediments, along with any organisms that happen to be within the tube diameter. Grabs (Figure 16.1c–e) take a "bite" out of the sea floor to capture sediments and any organisms contained within that sample, but this sampling technique typically results in some degree of disturbance to the sediment layering. Whereas corers and grabs sample vertically, dredges are pulled along the bottom of the ocean floor to scrape-up sediments and organisms into a trap. These samples offer marine biologists a snapshot in time of the sea floor and allow them to conduct studies of current biodiversity and ecology. Marine scientists typically focus on the organisms and often remove the sediment enclosing the organisms after a cursory study. However, the sediment is more than the "tomb" for the dead organisms, it is also the physical environment in which those organisms lived and interacted. To determine information about those environments, geologists analyze the sediment of ancient deposits in the same way that modern biologists analyze recent samples of sea floor (Figure 16.1f).

Due to the extended length of the process of sea-level change, a historical record must be established to determine events and trends of rise or fall. Within the sea floor, sediments are deposited vertically. This stratigraphy provides the historical context because, according to the Principle of Superposition in geology, older sediments are located at the bottom of a sediment pile, and the sediments become progressively younger closer to the surface.

The premise of the activity presented here is to use fossiliferous deposits from past sea-level changes as a model for the current phenomenon and demonstrate that ancient sea-level stands can be studied much like we study the modern ocean. Additionally, preservice teachers learn that the current accelerated sea-level shift, regardless of its cause, is the next phase within a progressive process that began long before humans began to affect the rate of change. This activity is used with Earth and planetary science preservice teachers, most of whom intend to teach middle school, and marine science majors in landlocked Tennessee.

Activity Version 1: Comparing Modern and Ancient Oceans

(Note: Text in italics represents commentary information for the reader relating this part of the activity to specific *NGSS* pedagogy.)

To prevent *a priori* concepts from influencing the outcome of the activity, I begin without providing any previous lecture or information regarding sea-level change history or mechanisms. I model the activity as "open-ended inquiry" and let the preservice teachers try to find the purpose of the activity for themselves. I provide locally collected, well-lithified fossiliferous marine rocks as one of the sea-floor samples to study (e.g., Paleozoic 400-million-year-old Ross Formation but any local sample will work), samples of lightly consolidated ancient marine sediments (e.g., Mesozoic 70-million-year-old Coon Creek Formation; see Figure 16.2a, p. 138), and samples of completely unconsolidated marine sediments from coastal Virginia that are only a thousand to a couple of million years old (e.g., Eastover Formation, Figure 16.2b), along with a modern ocean-grab sample collected from the Gulf of Mexico (Figure 16.1d, e). I inform the preservice teachers of the age and sampling location for all samples,

Figure 16.1. Types of Bottom Sampling Devices Used by Oceanographers to Sample Sea-Floor Sediments and Collect Organisms for Study

(a)　Tube-shaped core being extruded shipboard by the author

(b)　Extruded core split to show internal stratigraphy of the sediment

(c)　Peterson grab being used to "bite" a section of sea floor

(d)　Peterson grab sample opened on boat deck for study; deeper layers are to the outside and the center darker region is the sea floor nepheloid layer

(e)　Sample retrieved by a dredge across the sea floor; fine-grained material has been washed out, leaving shell hash

(f)　Dried sediment sample for microscope study after washing to remove mud

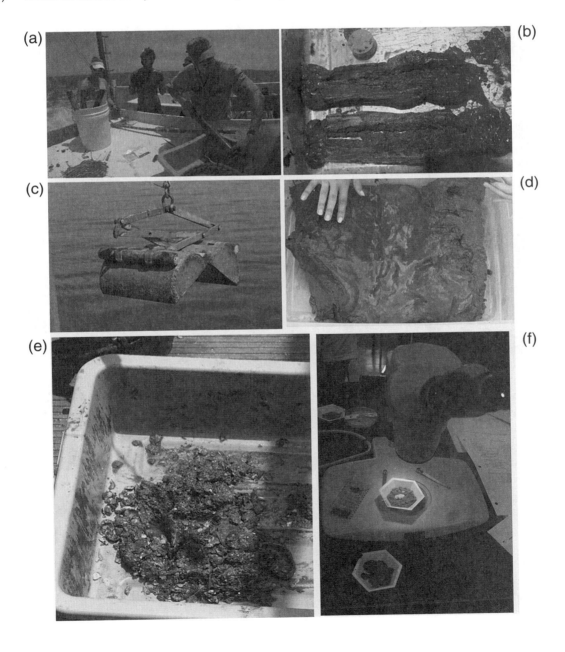

Figure 16.2. Ancient Sea-Floor Sediments

(a) Coon Creek Formation (72 million years old) sediment as collected from outcrop in dry state; note fossil content

(b) Eastover Formation (~10–5 million years old) from Virginia showing sand and fossil content

(c) Reconstituted wet Coon Creek sediment resembling a modern Peterson grab

(d) Reconstituted Eastover Formation sediment and fossils resembling a Peterson grab

(a) (b)
(c) (d)

crabs, shrimp, turtles, sharks) and ones that are now extinct (mosasaurs and plesiosaurs). To show contrast in regional geographic distribution, the Cenozoic sample (Figure 16.2b, d) is from Virginia, but it could easily also have also been from Tennessee. All the samples are dry when distributed (Figure 16.2a, b), including the Gulf of Mexico modern grab sample. It is useful for the preservice teachers to see that modern ocean samples can still be interpreted despite being dried. In fact, once modern oceanographers have recorded the chemical parameters of the ocean water, they usuallly remove the water before studying the organisms and the sediments. Geologists studying ancient oceans do not have access to the ocean water, but they can still analyze its properties using the proxies of sediments and organisms.

Prior to being distributed, the ancient marine and coastal sediment samples have been processed to resemble a modern sea floor grab sample by being partly or wholly "de-lithified."

This is accomplished by first adding water and then redrying the samples to produce the same type of sediment sample that would have been encountered by a "paleomarine geologist" who had traveled back in time to conduct a "Peterson paleograb." The preservice teachers can then study the material as they would a modern marine sediment (see Figure 16.3). For each sample, the preservice teachers are asked to complete the following tasks:

Observe and gather data to identify the *rocks and sediments* (composition and physical features) in each sample using the charts provided. Then, without making interpretations and relying only on observations, answer the following questions:

- What minerals can you identify (e.g., quartz, calcareous bioclasts, lithics)? Estimate percentages.
- What is the rock type? [Omit in version 2]
- What is the sediment type? Identify the sample as lithogenous, biogenic, hydrogenous, or cosmogenous (or a combination).
- What are the grain characteristics (grain size, sorting, rounding, etc.)?
- Name the sediment using one of the classification schemes discussed in class.

which allows them to see the time component from the beginning of the activity.

The Paleozoic sample contains organisms very different from modern marine organisms and represents a once tropical marine environment for Tennessee. This is due, primarily, to a combination of long-term evolutionary changes in organisms, extinction events, and preservation characteristics. The Mesozoic sample (Figure 16.2a, c) represents "dinosaur time" in Tennessee, and it shows that most of Tennessee was then an ocean similar to the modern Gulf of Mexico. Additionally, this sample contains both organisms similar to those found in modern oceans (clams, snails,

Figure 16.3. Laboratory Analysis of Both Fossil and Modern "Peterson Paleograbs"

Students analyze grain composition, texture, and organisms present in the sediment to determine environments and establish sea-level changes.

Observe and gather data to identify the *organisms* present in each sample using the charts provided, and then answer the following questions:

- What taxa are present?
- Are the organisms whole or fragmented?
- What is the mineral composition of the organism?
- Look up the geologic range of the organisms you identified on the range charts provided (or on the internet). What is the geologic age of this sample? [Omit in Version 2]

Observe and gather data regarding *sedimentary structures* in the sample using the identification charts provided. For example, do you see graded bedding, cross bedding, mud cracks, ripple marks, raindrop impressions, and so on? *At this point, preservice teacher–driven questioning of the initial data observed above takes center stage as participants try to make sense of their phenomena data, which invariably leads to additional questioning. For example, cause and effect relationships are sought (e.g., DCI MS-ESS5 crosscutting relationships connected to DCI MS-ESS3 cause and effect). Participants are instructed to use their phenomenon sample data to devise and answer inquiries relating to the environment and processes that may have resulted in the samples. They are also asked to cite evidence from their findings to support their responses. In general, the questions will be generated by the preservice teachers, with the instructor providing guidance if no suitable inquiries are developed. It is important to impress on the preservice teachers that observational data and interpretations of the data are to be kept separate in order to avoid circular reasoning.*

- Do the sediments or rock represent a terrestrial or a marine environment?
- Do the organisms represent terrestrial or marine organisms?
- Do the sedimentary structures represent terrestrial or marine settings?
- What observations can you make about energy level in the environment based on the sediments and the condition of the organisms?
- What conclusions can you make about the depth of water (e.g., below wave-base, no sunlight, etc.)?
- Look up the geologic range of the organisms you identified. Which ages were marine and which ages were terrestrial? Place the samples in their correct order from oldest to youngest. [Omit in Version 2]

Synthesis of Interacting Systems. *This is a second phase of student-generated questioning that occurs on a higher plane of comprehension and connects both time and space to sea-level phenomena in order to emphasize connections between concepts.* Now use your data and initial interpretations from each sample and apply them to the questions below. [Note: In Version 2, any questions participants haven't already posed are presented after the participants have been informed that one or more of their samples is actually a fossil deposit. Then, they are asked to make comparisons with the modern and ancient Peterson grab sample.]

- Compare the fossil deposits to the modern Gulf of Mexico deposits. How are they alike, and how do they differ? Compare both the sediments and the organisms. [Version 2 wording: Compare the grab samples. How are they alike, and how do they differ? Compare both sediments and organisms.]

- The area where the sample was collected is currently terrestrial and miles inland. Has sea level changed from the time of each sample to today? Would sea level have had to rise or drop to account for the fossil deposit? How does your data support your interpretation? [Omit in Version 2]

- Use your data to hypothesize how many times sea level may have changed over this time frame. [Omit in Version 2]

- Patterns similar to those observed in these samples have been found globally. What does this indicate about the processes responsible for these patterns? Are these processes local or global in nature? [Omit in Version 2]

- You only looked at a few samples of the total number of samples that you could have analyzed from the same region. In the vast majority of places, the pattern you documented dominates (i.e., marine or terrestrial). What conclusions can you make about whether or not the "modern" sea-level position is the norm for your area? What implications does this observation have for our current ideas of global change, sea-level change, and the way "things should be"?

It is important to note that all the samples of past oceans are from different sea-level incursions. For this reason, it is necessary to compare the dates of the samples used in this activity to the sea-level curve that was provided. It is also necessary for all the samples to have been collected within the same general inland region. Furthermore, the samples must be representative of the normal and dominant stratigraphy within the state. Oceans are ubiquitous and dominate most of geologic time, and the primary goals of this activity are to demonstrate that the well-lithified rocks are indeed marine as indicated by the marine fossils, that this means that the local area was covered by an ancient sea, and that there is a local record of sea-level change available everywhere regardless of the modern sea level. Well-lithified rocky material is generally very old geologically (tens to hundreds of millions of years old), and as such is often mistakenly considered too old to be relevant to modern sea-level change. Nevertheless, it does demonstrate that sea-level change is a real phenomenon, that it is a normal process on Earth, and that it can be studied using scientific means regardless of location.

Although any geologic time period can be used for this activity, sediments from the Cenozoic Era (66 million years old to today) are best because the fauna more closely resembles modern sea-floor fauna due to the occurrence of fewer extinction events within that time frame. However, older sediments make interesting comparisons too, and extensions of this activity can be used to introduce additional concepts related to organic evolution, extinctions, and Earth history.

The primary take-home lesson of this activity is that multiple oceans have occupied West Tennessee over the course of millions of years. Sea-level change is not a new phenomenon but rather a normal global-scale process of the planet, and current sea level is well below these numerous earlier sea-level events. Additionally, if most of the geologic history of an area is represented by marine histories, then not only is a higher sea level the norm for Earth but also the current shift toward a higher sea level should be viewed as returning to the norm, not a deviation from the norm. Ultimately, the participants realize that the human effect on sea level affects the rate of change rather than the existence of the phenomenon.

Activity Version 2: The Reconstituted Ocean

This version of the activity is usually most useful with students who have little or no Earth science background, and it works particularly well with preservice teachers with more extensive biological science training but have spent little time considering the role that time plays in science.

The second version of the activity begins, as the first did, with little background information other than a description of the three bottom sampling methods (cores, grabs, and dredges). The preservice teachers are asked to compare two Peterson grab samples without any knowledge of location, water depth, and so on. Instead, it is their job to infer that information. The participants are provided with one or two modern sediment grabs and one or two ancient ocean "grabs" from Mesozoic and Cenozoic seas at the same time, and both samples are wet when presented (Figure 16.2c, d). In this version, I do not give out information regarding the geologic age of the sediments. Most participants will assume that all samples are modern, although more astute observers may note that the color is absent from the shells of some deposits and try to develop elaborate hypotheses to explain this observation.

All of the sediments must be wet enough to look as though they were recently collected from an actual Peterson grab on a boat trip. The modern sample contains its original saltwater from collection, and the ancient samples have aquarium saltwater, real ocean saltwater, or sometimes just freshwater added to allow them to become de-lithified into a wet sediment, which then becomes a reconstituted seafloor. Thus, all the sediments appear to represent modern grab samples from different sea-floor settings.

The preservice teachers are next asked to describe the sediments and organisms and compare the environments, which are typical activities for any biology or ecology class. Once the preservice teachers have attempted to generate their own questions, the activity uses the same guiding questions listed in the first version but with additional biological questions pertaining to shell morphology and other aspects of marine biology. Even if they note differences in biota, water clarity, sediment type, and so on, most participants will continue to assume that the samples are from different areas within the same modern ocean. Invariably, all interpretations are ecological in nature and assume the same time frame of coexistence. At this point, sea-level change is not considered a viable hypothesis.

Later, after the preservice teachers have generated their initial set of questions and completed a preliminary analysis (using what I call their "biologically myopic eyes") to develop a list of the environmental and ecological differences between the sites, they are informed that some of the samples were collected from a nearby site on dry land, and that these samples are actually reconstituted ancient sea floor. This is usually a eureka moment of discovery, and they often protest that it was unfair to withhold this information. They immediately

begin to postulate about which samples are modern and which are not. At that point, the preservice teachers are provided with the guiding questions that had been omitted from the first version. They are then asked to reinterpret their original data set with the knowledge that the samples are from two different oceans from two different geologic periods in the same general area. Invariably, change in sea level is called upon to explain the distribution of marine sediments. My overall goal in this version of the activity is for the preservice teachers to be able to visualize for themselves that their now terrestrial world still contains the evidence of its oceanic past when sea level was much higher than today.

Information for Students

Fossiliferous marine sediments are found in nearly all states and are routinely studied by geologists and paleontologists for the purposes of reconstructing ancient environments and sea-level changes, so material is plentiful and readily available, as are detailed published analyses of them. Obtaining fossiliferous sediment samples is the first challenge for the teacher but can be accomplished with a little local research. The best resources for obtaining fossiliferous sediment samples are (1) geology departments at nearby universities, (2) state geological survey offices and museums, (3) local amateur rock hound organizations, (4) the websites of professional Earth science societies with education arms (e.g. Paleontological Society, Fossil Project), and professional science suppliers like Wards. Once obtained, the ancient ocean sediments must be reconstituted ahead of time.

Mineral, rock, sediment, and organism identification guides are typically found in most Earth science textbooks, and the books in the Teacher-Friendly Guides series (e.g., Ansley 2000; Swaby and et al. 2016) are particularly useful resources. If taxonomic identification is one of the learning targets, more detailed identification guides such as the Peterson Field Guides (PFG) or the National Audubon Society Field Guides can be used. Local fossil identification guides are available from the sources mentioned in the preceding paragraph as well as in any basic college geology, historical geology, or paleontology textbook or lab manual (e.g., Levin 2013; Prothero and Dott 2010; Ritter and Petersen 2011). These resources also contain geologic time scales, sea-level change diagrams, and ocean zonation diagrams that can be useful for preparing handouts or graphics, and most can be found at local libraries or online.

Conclusion

As faculty responsible for undergraduate education, we have learned a lot from using the *NGSS* phenomena approach while teaching preservice middle school educators to incorporate student-collected natural phenomena into their teaching of sea-level change and its associated issues. The *NGSS* emphasize crosscutting concepts with respect to patterns, cause and effect, scale and proportion, quantity, systems approaches, energy within systems, structure and function within systems, and stability and change within systems. The geosciences emphasize the systems approach to studying the Earth, adding the component of time throughout history. Earth science time scales extend well beyond the human experience, and the rock record contains the physical evidence, or data, needed to comprehend issues related to large-scale systems and global change over realistic time scales.

The fact that Earth science phenomena are not confined to laboratory modeling experiments, can be place-based as well as global in scope, integrate well with all other disciplines of science, and use technology and mathematics extensively while remaining largely experiential in nature makes them ideal for use in *NGSS*-aligned curricula. These characteristics allow preservice teachers to easily and readily relate directly to Earth phenomena in a personal and meaningful way, thus creating an informed perspective on these issues. In my experience, undergraduate preservice teachers often perceive Earth science phenomena such as sea-level change and volcanic eruptions as extreme events even though they are actually occurring every day. Impending doom tends to attract attention, and this perception of geologic events as extreme can be used to pique a natural motivation to question and a desire to know more.

In the process of educating educators about using the *NGSS* to teach about Earth science phenomena, especially controversial phenomena that span long periods of geologic time and operate over global scales that affect individuals and society, we have learned that the *NGSS* teach perspective and self-discovery. Just as no one can understand the plot of a movie by only seeing the last minute or even by only being told about it, global climate change cannot be fully understood when only factors from the last few centuries are taken into consideration. Current understanding of global change and sea-level issues is incomplete and often completely erroneous due to a public that is poorly educated in Earth science. Practicing geoscientists have always accepted the idea that change, such as that represented by sea-level change, occurs on a global and geologic scale (e.g., Breslyn et al. 2016). However, it is more difficult to convey this knowledge to the general public, both because Earth science is the scientific discipline that is taught the least frequently in public schools and because the instructional methods used when it *is* taught often only provide a superficial level of understanding. While it is true that the controversial nature of the core ideas present challenges in a public school setting, the *NGSS* can help teachers deal tactfully with these issues by encouraging students to arrive at their own conclusions through analysis of readily available local data.

Historical geology courses have used geologic evolution as their primary organizing theme since they were first taught in the 1700s. Whereas most history is generally something that is memorized rather than experienced, the long-term historical results of sea-level change caused by global change still exist, and these changes are evident everywhere those oceans were located historically, not only at modern coasts. This historical evidence makes sea-level change as represented in the geologic record an ideal phenomenon for direct student data collection that lends itself to the construction of explanatory arguments supported with evidence. The effects of sea-level change over the ages has been preserved, allowing evidence of the phenomenon to be seen, touched, and explained by students through the process of phenomenological inquiry. Most important, pedagogical methods that align with the *NGSS* allow students to see evidence of sea-level change locally and to relate to the phenomenon personally.

Rock and sediment strata archive the environments of the past as well as changes in those environments. The well-known geologic uniformitarian principle espoused by Sir Charles Lyell, "the present is the key to the past," may apply well to the study of geohistorical events, but in reality the present is the *result* of the past—which means the past is our best predictor of the various possible futures of our planet.

References

Ansley, J. E. 2000. *The teacher-friendly guide to the geology of the northeastern US*. Ithaca, NY: Paleontological Research Institution.

Baumgartner, E., L. Phillips, and E. K. Maynard. 2008. Building ocean literacy through science and art. *Currents* 24 (3): 18–23.

Breslyn, W., J. R. McGinnis, R. C. McDonald, and E. Hestness. 2016. Developing a learning progression for sea-level rise, a major impact of climate change. *Journal of Research in Science Teaching* 53 (10): 1471–1499.

Coe, A. L. 2003. *The sedimentary record of sea-level change.* Cambridge, UK: Cambridge University Press.

Davis Jr., R. A. 2011. *Sea-level change in the Gulf of Mexico.* College Station, TX: Texas A&M Press.

Evans, D. 2016. Survey reveals challenges with teaching climate change. NSTA blog. *www. nstacommunities.org/blog/2016/02/12/ survey-reveals-challenges-with-teaching-climate-change/*

Hestness, E., R. C. McDonald, W. Breslyn, J. R. McGinnis, and C. Mouza. 2014. Science teacher professional development in climate change education informed by the *Next Generation Science Standards. Journal of Geoscience Education* 62 (3): 319–329.

Holthuis, N., R. Lotan, J. Saltzman, M. Mastrandrea, and A. Wild. 2014. Supporting and understanding students' epistemological discourse about climate change: *Journal of Geoscience Education* 62 (3):174–387.

Levin, H. R. 2013. *Earth through time.* 8th ed. New York: Wiley Press.

Munsart, C. A. 1997. *American history through Earth sciences.* Santa Barbara, CA: ABC-CLIO.

NGSS Lead States. 2013. *Next Generation Science Standards: For states, by states.* Washington, DC: National Academies Press. *www.nextgenscience.org/ next-generation-science-standards.*

National Research Council (NRC). 1996. *National Science Education Standards.* Washington, DC: National Academies Press.

National Research Council (NRC). 2000. *Inquiry and the National Science Education Standards: A guide for teaching and learning.* Washington, DC: National Academies Press.

Plutzer, E., M. McCaffrey, A. L. Hannah, J. Rosenau, M. Berbeco, and A. Reid. 2016. Climate confusion among U.S. teachers. *Science* 351 (6274): 664–665.

Prothero, D. R., and R. H. Dott Jr. 2010. *Evolution of Earth.* 8th ed. New York: McGraw-Hill.

Ritter, S. M., and M. Peterson. 2011. *Interpreting Earth history: A manual in historical geology.* 7th ed. Long Gove, IL: Waveland Press.

Robertson, W. H., and A. C. Barbosa. 2015. Global climate change and the need for relevant curriculum. *International Journal of Learning, Teaching and Educational Research* 10 (1): 35–44.

Smith, L. K., M. Barber, L. Duguay, and L. Whitley. 2012. Using the ocean literacy principles to connect inland audiences to the global oceans. *Currents* 28 (2): 2–7.

Strang, C., K. DiRanna, and J. Topps. 2010. Developing the ideas of ocean literacy using conceptual flow diagrams. *Currents Special Report* 3: 27–62.

Swaby, A. N., M. D. Lucas, and R. M. Ross. 2016. *The teacher-friendly guide to the Earth science of the southeastern U.S.* 2nd ed. Ithaca, NY: Paleontological Research Institution.

Trendell Nation, M., A. Feldman, and P. Wang. 2015. A rising tide. *The Science Teacher* 82 (6): 34–40.

Michael A. Gibson *is a professor of geology at the University of Tennessee-Martin with specialties in paleoecology and marine geology. He is the author of more than 45 journal articles and book chapters and more than 200 professional presentations. He is an associate curator for the Pink Palace Museum in Memphis, Tennessee, and of the Coon Creek Science Center. During summers, he teaches courses in marine geology and sea-level change at the Dauphin Island Sea Lab in Alabama. He has received more than 30 awards, including the Neil Miner Award from the National Association of Geoscience Teachers, and he is a University of Tennessee Alumni Association distinguished service professor. Gibson can be contacted by e-mail at* mgibson@utm.edu.

CHAPTER 17

Challenges in Undergraduate Science Teaching for Prospective Teachers

Jay B. Labov

In this chapter, we begin with a focus on the ways in which recent K–12 educational initiatives, such as the *Next Generation Science Standards* (*NGSS*; NGSS Lead States 2013) and the recently implemented restructuring of Advanced Placement science courses, are likely to affect teaching and learning of STEM subjects in the future. We also look at the ways in which college and university faculty might better prepare preservice teachers of STEM to meet these new expectations. Finally, we discuss various aspects of education policy and the ways in which deepening policy understanding among individual faculty and STEM departments can improve the education of prospective teachers.

Not long ago, a college chemistry professor grew angry with the way her daughter's high school chemistry class was being taught. She made an appointment to meet with the teacher and marched with righteous indignation into the classroom—only to discover that the teacher was one of her former students.

—Yates 1995, p. 8b

Although the above scenario was described almost 20 years ago and cited in a National Research Council report several years later (NRC 1999), it still rings true today. College and university faculty frequently decry a perceived lack of both content knowledge and analytical skills among undergraduate STEM students, and they then blame K–12 teachers for failing to properly prepare those students.

Although the cultivation of STEM-related knowledge and reasoning is undeniably critical during primary and secondary education, postsecondary faculty, especially in the STEM disciplines, often fail to recognize that adequate preparation of K–12 students and teachers is a highly complex and evolving process. Although higher education is a critical component of that process, its contributions to improving K–12 education are often lacking. Teachers who are underprepared or who fail to motivate students can hinder learning and academic achievement in students for years (see National Academies of Sciences, Engineering, and Medicine 2015 for a review of the literature on this topic). By contrast, teachers who are well versed in their subject matter and can inspire students to learn and achieve are often able to overcome these deficits and make up ground.

Many college and university faculty may not know that K–12 education in STEM has undergone profound changes in policy and emphasis. Among these changes was the development, publication, and attempted implementation of numerous sets of standards for STEM subjects (Table 17.1, p. 146). As these standards have evolved, they have increasingly emphasized engaging students actively with subject matter, integrating engineering and technology into the teaching and learning of science and mathematics, and emphasizing crosscutting themes within and across grade levels and between different sets of standards (e.g., NRC 2012a; NGSS Lead States 2013; CCSS 2010).

Another recent initiative was the fundamental restructuring of Advanced Placement courses in biology, chemistry, and physics that took place during the past four years (College Board 2011, 2014, 2015; NRC 2002). Each of these subjects now emphasizes several high-level concepts and enduring understandings. Teachers are encouraged to "cover" fewer topics and explore those topics in much greater depth. Additionally, expectations for scientific competencies that are common to the three disciplines have been identified and emphasized.

Change has also occurred with the application of research conclusions to instruction, leading to improvements in teaching and learning (e.g., NRC 2000, 2012a; Handelsman et al. 2007; Wiggins and McTighe 2015; Kober 2015). This evidence base was largely responsible for driving changes to more recent sets of standards and determining which methodologies are currently recommended for frequent use in K–12 instruction.

At the federal level, policy updates have been a matter of continuous debate since the early 2000s with implementation of the No Child Left Behind Act (NCLB), which has had a large, unanticipated, and sometimes negative effect on precollege STEM education. The reform required every child to be tested in English and mathematics each year from grades 3–8, and schools were expected to show adequate yearly progress (AYP) toward proficiency for all students; however, proficiency was determined by individual

Table 17.1. K–12 Standards for STEM Subjects, 1993–Present

Standards	Issuing Organization	More Information
Benchmarks for Science Literacy (1993)	American Association for the Advancement of Science	*www.project2061.org/publications/bsl/default.htm*
National Science Education Standards (1996)	National Research Council	*www.nap.edu/catalog/4962/national-science-education-standards*
Principles and Standards for School Mathematics (2000)	National Council of Teachers of Mathematics	*www.nctm.org/standards*
Standards for Technological Literacy (2000); *Content for the Study of Technology* (2002); *Content for the Study of Technology* (2007)	International Technology Education Association	*www.iteea.org/39197.aspx*
Advancing Excellence in Technological Literacy: Student Assessment, Professional Development, and Program Standards (2003)	International Technology Education Association	*www.iteea.org/42523.aspx*
Common Core State Standards, Mathematics and English Language Arts (2010)	Common Core Standards Initiative	*www.corestandards.org*
A Framework for K–12 Science Education: Practices, Crosscutting Concepts, and Core Ideas (2012)	National Research Council	*www.nap.edu/catalog/13165*
Next Generation Science Standards (2013)	NGSS Lead States	*www.nextgenscience.org*

states, resulting in vast differences in performance expectations for students from different states. Science was not subjected to the AYP requirement and was to be tested once at each of the three educational stages (elementary, middle, and high school). Various sanctions were to be imposed on schools deemed to be failing under NCLB. These high-stakes tests consisted largely of multiple-choice questions that assessed a low level of cognitive understanding, and many in the mathematics and scientific education communities worried that the necessity of teaching to these tests was undermining the goals of the standards.

The Every Student Succeeds Act (ESSA), a revision of NCLB signed into law in 2015, returns more authority to states by allowing them to set their own methods of measuring school and student success by means of state-level accountability systems and assessments. This revised system is more amenable to the promotion of STEM skills development because it allows states to identify science and other subjects beyond reading and mathematics as areas of success. Gomoran (2016) considers ESSA much more accommodating to advances in knowledge about STEM learning than NCLB was.

Although the requirements for becoming a teacher of STEM vary widely across states (see NRC 2010), increasingly states are requiring future teachers to major in a discipline rather than in education. While this requirement historically has been the norm for teachers with certification in grades 9–12, it has now been extended in some states to cover the middle grades as well. Acquiring a degree in a STEM field is not yet a widespread prerequisite for teachers in the elementary grades, since they commonly teach across content areas, although these teachers will have a great influence on the ways in which upcoming generations will view STEM.

Roles and Responsibilities of Higher Education Faculty, Departments, and Institutions

RECRUIT AND TRAIN 100,000 GREAT STEM TEACHERS OVER THE NEXT DECADE WHO ARE ABLE TO PREPARE AND INSPIRE STUDENTS: *The most important factor in ensuring excellence is great STEM teachers, with both deep content knowledge in STEM subjects and mastery of the pedagogical skills required to teach these subjects well. … The federal government should* set a goal of ensuring over the next decade the recruitment, preparation, and induction support of at least 100,000 new STEM middle and high school teachers who have strong majors in STEM fields and strong content-specific pedagogical preparation, by providing vigorous support for programs designed to produce such teachers.

—President's Council of Advisors on Science and Technology 2010, p. viii

All of us who teach science at the college level need to face the fact that our teaching sets the standard for science education at all lower levels. Thus, for example, if professors only lecture to passive students, aiming to attain maximum coverage of the vast subject of biology in their introductory biology classes, college teaching will remain the major obstacle in the path of science education reform. Research carried out in the past few decades conclusively demonstrates that active learning can be incorporated effectively into even large lecture classes. As scientists, we can and we must do better.

—Alberts 2016, p. 5

Many faculty do not fully appreciate the ways in which the evolution of K–12 STEM education and education policies are likely to influence them and their departments and institutions in the future. Given the current structures, hierarchies, and competing incentives for success in higher education, STEM faculty also may not fully appreciate either their critical roles or their important responsibilities in ensuring that K–12 teachers have the knowledge, skills, and attitudes to be successful in their own classrooms while educating an increasingly diverse population of students.

Despite the fact that nearly all future STEM teachers must enroll in discipline-based courses at least through the introductory level (and, as previously noted, many must major in a STEM discipline), many postsecondary instructors still view primary responsibility for the preparation of K–12 teachers of STEM as lying with the school or college of education rather than with themselves or their departments, or as a shared mission between the STEM disciplines and the school of education.

Faculty also may not fully appreciate that the pedagogies they employ to teach STEM subjects can set a crucial example for preservice teachers. Too often, college-level STEM courses are large lectures that cover a broad range of topics without providing the means for these teaching candidates to develop a deeper understanding of content and concepts. Postsecondary survey courses in STEM fields also often obscure the relevance of topics and connections throughout the course (or beyond). As a result, teaching candidates may receive mixed and conflicted messages about appropriate ways to teach STEM subjects (e.g., Kober 2015, NRC 2000).

Many students who enter college intending to major in a STEM discipline also may be considering a K–12 teaching pathway. However, research shows that some 60 percent of these students nationally opt for other majors after taking the introductory course sequence in their preferred discipline. When asked why they are transferring out of STEM, many report that their courses were boring, irrelevant to their own interests, and lacked a clear connection to real-world problems (PCAST 2012, Seymour and Hewitt 1997). Most of these students were not unqualified for careers in STEM, having passed the courses in which they had enrolled. Because many students do not know when they enter college whether they are committed to pursuing a teaching career path, the proportion who leave STEM at that point cannot be determined.

Those who do continue on a path toward teaching tend to use experience in the STEM courses they take to determine the types of information and skills that are most important. When testing large numbers of students in survey courses, postsecondary instructors often rely on multiple-choice tests that are easier to grade than exams requiring multipart responses or essays. However, if these tests also emphasize lower-level knowledge and skills such as factual recall or definitions, they can give preservice teachers the wrong idea of what they should assess. Fortunately, there exists a large body of research about testing methods that effectively assess higher-order thinking skills in large survey courses that enroll preservice teachers (e.g., Dirks et al. 2014).

We want to prepare 100,000 new teachers in the fields of science and technology and engineering and math. In fact, to every young person listening tonight who's contemplating their career choice: If you want to make a difference in the life of our nation, if you want to make a difference in the life of a child, become a teacher. Your country needs you.

—President Barack Obama,
State of the Union Address, 2011

One way for postsecondary instructors to reinforce the importance of high-quality teaching is to model effective pedagogies in their courses. Handelsman et al. (2007), Kober (2015), and Bradforth (2015) all offer helpful, evidence-based guidance for changing pedagogies to incorporate a variety of active-learning techniques in STEM courses, including courses with large enrollments. Campus teaching and learning centers can also provide valuable advice and real-time feedback to faculty. Many STEM-related disciplinary societies offer workshops on effective undergraduate teaching (e.g., Hilborn 2013). Other national efforts include the Summer Institutes on Scientific Teaching (formerly the National Academies Summer Institutes on Undergraduate Education; Pfund et al. 2009; Labov and Young, 2013; see *www.summerinstitutes.org*), the Science Education for New Civic Engagements and Responsibilities (SENCER) Summer Institutes (see *www.sencer.net/symposia*), and the Undergraduate STEM Initiative of the Association of American Universities (see *https://stemedhub.org/groups/aau*).

Graduate students and postdoctoral fellows who plan to pursue careers that include university teaching should also be encouraged to learn about and employ new pedagogies. National programs such as The Preparing Future Faculty Initiative (see *www.preparing-faculty.org*) and Center for the Integration of Teaching and Learning (CIRTL) Network (see *www.cirtl.net*) focus on providing graduate students with the knowledge and resources necessary to do this.

Unfortunately, some instructors who serve as mentors to undergraduate students have been known to question or denigrate their best students' desires to become teachers, urging them to pursue additional education in graduate or professional schools instead. These conversations send powerful messages to teaching candidates that teaching is not a profession worth pursuing. Additionally, if faculty accept only students who are planning to pursue graduate education as apprentices in research programs or in course-based research experiences (National Academies of Sciences, Engineering, and Medicine 2015) and excludes prospective STEM teachers, the message sent is that K–12 teachers are not a part of or valued within the scientific community. The National Science Foundation's Research Experiences for Teachers initiative (available through several directorates) is one resource that provides support for current teachers to gain these kinds of invaluable hands-on research experiences in cutting-edge science and technology.

Examples of Roles of Departments and Institutions

Departments and institutions of higher education also can play critical roles in supporting undergraduates who have declared their intention to become K–12 teachers or who are contemplating doing so. Students who enter college wishing to become doctors or lawyers are often assigned (or have easy access to) advisers with expertise in these areas. But how many STEM departments or institutions offer similar, dedicated services to preservice teachers outside of the school of education? Offering preservice teachers access to advisers with specialized expertise and adopting comprehensive teacher preparation programs are ways in which institutions can demonstrate that K–12 teaching is valued, important, and supported. One such program is UTeach, which began at the University of Texas at Austin, and is now, through support from the National Science and Math Initiative (see *www. nms.org/Programs/UTeachExpansionProgram.aspx*), being replicated on a national level.

In addition to offering teaching candidates a supportive program, it is essential that STEM and education faculty and departments communicate with one another about common goals and approaches that promote the *NGSS*. When communication is lacking, problems rooted in a perceived devaluation of K–12 teaching as a worthwhile career pathway likely will continue to be exacerbated, particularly from students' perspectives. It is important to recognize that most teachers of STEM in the elementary or middle school grades may not be required to take courses in STEM beyond the introductory level, so the strategies that STEM faculty model or fail to model in these courses is of utmost importance.

The challenges of preservice education are complex and require all components of the education system to recognize that their actions (or lack thereof) influence other components of the system. Readers of this volume will find a multitude of insights, perspectives, and specific approaches to improving education for future teachers of STEM.

References

Alberts, B. 2016. An addiction to education at all Levels. *Lasker Awards Stories*. *Cell* 167: 1–5.

Bradforth, S. E. 2015. University learning: Improve undergraduate science education. *Nature* 523 (7560): 282–284.

College Board. 2011. *AP Biology curriculum framework, 2012–2013.* New York: College Board. *www.media. collegeboard.com/digitalServices/pdf/ap/10b_2727_AP_ Biology_CF_WEB_110128.pdf*

College Board. 2014. *AP Chemistry course and exam description.* Rev. ed. New York:

College Board. *https://secure-media.collegeboard.org/ digitalServices/pdf/ap/ap-course-overviews/ap-chemistry-course-overview.pdf*

College Board. 2015. *AP Physics I: Algebra-Based and AP Physics II: Algebra-Based course and exam description, including the curriculum framework.* Rev. ed. *http://media. collegeboard.com/digitalServices/pdf/ap/ap-physics-1-2-course-and-exam-description.pdf.*

Dirks, C., M. P. Wenderoth, and M. Withers. 2014. *Assessment in the college science classroom.* New York: W. H. Freeman.

Gamoran, A. 2016. Will latest U.S. law lead to successful schools in STEM? *Science* 363: 209–211.

Handelsman, J., S. Miller, and C. Pfund. 2007. *Scientific teaching.* New York: W. H. Freeman.

Hilborn, R., ed. 2013. *The role of scientific societies in STEM faculty workshops.* Washington, DC: Council of Scientific Society Presidents. *http://www.aapt.org/ Conferences/newfaculty/upload/STEM_REPORT-2.pdf.*

Kober, N. 2015. *Reaching students: What research says about effective instruction in undergraduate science and engineering.* Washington, DC: National Academies Press.

Labov, J., and J. Young. 2013. National Academies Summer Institutes on undergraduate education in biology. *The role of scientific societies in STEM faculty workshops,* R. Hilborn, ed., 21–24. Washington, DC: Council of Scientific Society Presidents.

National Academies of Sciences, Engineering, and Medicine. 2015. *Science Teachers' Learning: Enhancing Opportunities, Creating Supportive Contexts.* Washington, DC: National Academies Press.

National Governors Association Center for Best Practices and Council of Chief State School Officers (NGAC and CCSSO). 2010. *Common core state standards.* Washington DC: NGAC and CCSSO.

National Research Council (NRC). 1999. *Transforming undergraduate education in science, mathematics, engineering, and technology.* Washington, DC: National Academies Press.

National Research Council (NRC). 2000. *How people learn: Brain, mind, experience, and school.* Expanded ed. Washington, DC: National Academies Press.

National Research Council (NRC). 2002. *Learning and understanding: Improving advanced study of mathematics and science in U.S. high schools.* Washington, DC: National Academies Press.

National Research Council (NRC). 2010. *Preparing teachers: Building evidence for sound Policy.* Washington, DC: National Academies Press.

National Research Council (NRC). 2012a. *A framework for K–12 science education: Practices, crosscutting concepts, and core ideas.* Washington, DC: National Academies Press.

National Research Council (NRC). 2012b. *Discipline-based education research: Understanding and improving learning in undergraduate science and engineering.* Washington, DC: National Academies Press.

National Research Council (NRC). 2015. *Integrating discovery-based research into the undergraduate curriculum: Report of a convocation.* Washington, DC: National Academies Press.

NGSS Lead States. 2013. *Next Generation Science Standards: For states, by states.* Washington, DC: National Academies Press. *www.nextgenscience.org/next-generation-science-standards.*

Pfund, C., S. Miller, K. Brenner, P. Bruns, A. Chang, D. Ebert-May, et al. 2009. Summer institute to improve biology education at research universities. *Science* 324: 470–471.

President's Council of Advisors for Science and Technology (PCAST). 2010. *Report to the president: Prepare and inspire: K–12 education in science, technology, engineering, and math (STEM) for America's future.* Washington, DC: PCAST.

President's Council of Advisors for Science and Technology (PCAST). 2012. *Report to the president: Engage to excel: Producing one million additional college graduates with degrees in science, technology, engineering, and mathematics.* Washington, DC: PCAST.

Seymour, E., and N. M. Hewitt. 1997. *Talking about leaving: Why undergraduates leave the sciences.* Boulder, CO: Westview.

Slater, S., T. Slater, and J. M. Bailey. 2010. *Discipline-based science education research: A scientist's guide.* New York: W. H. Freeman.

Wiggins, G., and J. McTighe. 2015. *Understanding by design.* Expanded 2nd ed. Alexandria, VA: ASCD.

Yates, A. *Denver Post.* 1995. Higher Education Has a Link to Real Reform at the K–12 level. April 29.

Jay B. Labov *is senior adviser for education and communication at the National Academies of Sciences, Engineering, and Medicine. He has directed or contributed to some 30 reports on K–12 and undergraduate, teacher, and international education. He is currently serving as director of the Academies' Teacher Advisory Council, and he has also directed the Committee on Revising Science and Creationism. Additionally, he continuously oversees efforts to confront challenges to teaching evolution in the nation's public schools. He has received a lifetime honorary membership of the National Association of Biology Teachers, an education fellowship from the American Association for the Advancement of Science, and NSTA's Distinguished Service to Science Education Award. Labov can be reached by e-mail at jlabov@nas.edu.*

5

Epilogue:
Three-Dimensional Instruction Beyond the Classroom

CHAPTER 18

Harnessing the Business Community and Other Entities to Support the Vision of the *NGSS*

Chih-Che Tai, Ryan Nivens, Laura Robertson, Karin Keith, Anant Godbole, and Jack Rhoton

In this chapter, we discuss the strategies that the East Tennessee State University Center of Excellence in Mathematics and Science Education and the Northeast Tennessee STEM Innovation Hub have used to leverage partnerships, networks, and community collaborations to help school districts implement the vision called for in *A Framework for K–12 Science Education* (*Framework*; NRC 2012) and the *Next Generation Science Standards* (*NGSS*; NGSS Lead States 2013). Successful partnerships involve a wide array of individuals and groups, including teachers and administrators, business leaders, policy makers, and other community groups. In this chapter, we discuss how to establish these partnerships, sustain active involvement in them, and use them to support instructional reform. Finally, we discuss the challenges of maintaining successful educational partnerships and the lessons that we've learned while addressing those challenges.

Introduction

Partnerships can be powerful tools for bringing about change, and businesses partner with schools to promote and support STEM activities at all levels of education. These partnerships will take on added meaning and importance as we engage business and industry, as well as other entities, in building innovative capacities to support the *NGSS*. The new standards present an unprecedented challenge by requiring districts to build systems for helping teachers to attain a deeper conceptual understanding of core science knowledge, develop a more informed perspective on science pedagogy, and enhance their awareness of the ways in which students learn science. This challenge requires districts to provide teachers with access to the expertise and materials necessary to actively engage all students in science and engineering practices while also emphasizing interdisciplinary connections to content areas such as mathematics and literacy.

Effective change begins at the community level, and local and regional STEM partnerships can expand a community's educational infrastructure to improve the quality of STEM teaching and learning. Society's increasing reliance on technology makes it more important than ever that education and industry work together to inform educational decisions and achieve common goals.

Partnerships between educational institutions and local businesses organize human resources, ideas, tools, and information from many to more fully invest educators in meeting the needs of local industry and vice versa. It was for these reasons that East Tennessee State University (ETSU), through its Center of Excellence in Mathematics and Science Education (CEMSE) and the ETSU Northeast Tennessee STEM Innovation Hub (NTSIH), has worked to create local and regional partnerships that serve school districts throughout Northeast Tennessee. These partnerships contribute support for practice-based research, professional development, informal STEM learning, technical assistance and expertise, learning communities, business and industry workforce needs, proposal writing, and other accomplishments and opportunities that would not be available otherwise. By describing the components of these partnerships, the activities that they make possible, and the benefits that result, we hope to encourage other communities to develop their own educational-business partnerships that enhance student experience and increase achievement.

A Strong Partnership Environment

Our work has benefited from our location in a vibrant and productive STEM community in Northeast Tennessee, which includes not only K–12 school districts, colleges, and a university but also large businesses such as Eastman Chemical Company, Wellmont Health System, and Domtar Paper Mill; science-rich institutions like the ETSU and General Shale Brick Natural History Museum, Hands On! Science Museum, and Bays Mountain Park and Planetarium; and foundations and smaller businesses such as Snap-on-Tools, Nuclear Fuel Services, Aerojet, LMR Plastics, ALO, and the Niswonger Foundation. All of these entities have worked in various partnerships to support STEM education in the region. One such partnership resulted in the creation of the Innovation Academy of Northeast Tennessee, a robust, groundbreaking school

that embraces STEM education to meet challenges of the global marketplace through innovation, collaboration, networking, and creative problem solving. These partnerships provide powerful, rich professional development for literally hundreds of math and science teachers and help them to tackle the intricate process of aligning the *Common Core State Standards (CCSS;* NGAC and CCSSO 2010) with the *NGSS*.

Our partnerships manifest inclusive leadership and offer job shadowing as well as learning opportunities for teachers. By establishing clear connections to STEM-related businesses and other types of STEM-learning environments, we have been able to link the region's classrooms to new sources of support and funding that are critical to the sustainability and long-term viability of the partnerships. As a result, our students learn through exploration and discovery as they are guided by teachers working jointly with scientists and other professionals from STEM fields. These experiences make STEM knowledge and skills more meaningful by connecting them to the real world.

The STEM Council

An important aspect of our work is rooted in the ETSU STEM Council, a sustainable, synchronized learning network composed of K–12 teachers and administrators, institutes of higher education, business and industry entities, and nonprofit associations. These partners collaborate to amplify and accelerate the reach of STEM education throughout the region. The mission of the council is to help students develop the intellectual capital to succeed in future careers and stimulate economic vitality for our region.

At its core, the STEM Council is a means of bringing interested parties together on a regular basis to build relationships and plan strategies. We have found that initiating conversations between the business and education communities can illuminate problems, such as knowledge and skill gaps common among entry-level STEM employees, and help create innovative delivery systems that address those problems. By capitalizing on the power and sway of individuals from each side of the partnership, we are able to create an atmosphere of trust and understanding that helps everyone to focus on a shared vision. All members of the partnership bring to the table expertise unique to their work, and their contributions combine to enact innovative solutions for achieving common goals.

Not all partners contribute in the same way. Large companies often have the capacity to invest significant financial resources, whereas smaller organizations may be more likely to contribute volunteers and mentors, technical assistance, and other forms of human resources. Some partners merely contribute by advocating for our work, being knowledgeable about who we are, and validating the need for qualified STEM workers. The business community recognizes that the quality of their workforce is closely related to the quality of local STEM education, and this relationship makes it a natural ally. Businesses and other entities provide educational assistance by creating environments for collaboration and feedback that result in ideas about ways to improve the teaching of STEM disciplines. Although the assistance provided by our partners takes many forms, the one factor they all have in common is that they enable educators to more effectively implement the shifts called for by the *Framework* and the *NGSS*.

Benefits of Partnerships

Cooperative partnerships allow us to improve STEM teaching and learning for students across Northeast Tennessee by implementing innovative strategies aligned with the *NGSS* and the science literacy components of the *CCSS*. The crosscutting concepts of the *NGSS* help science and English language arts (ELA) teachers to think differently about the best ways to teach science and literacy and to integrate those methods into their instruction. Research shows that language and literacy are learned best if they are embedded in meaningful contexts (Lightbrown and Spada 2006). Science knowledge and literacy skills are fundamental to our society and the work of future generations of professionals (Pearson, Moje, and Greenleaf 2010; Wallace 2004), and we have this in mind when helping inservice ELA, math, and science teachers to clarify the relationship between the crosscutting concepts of the *NGSS* and *CCSS* practices. For example, the *NGSS* call for students to define problems and ask questions as well as analyze and interpret data. Likewise, in English language arts, students are taught to identify problems described in text and ask questions about readings. In both disciplines, after formulating questions, students collect evidence, revise their thinking, and make predictions. Due to their interdisciplinary nature, these procedures and constructs are embedded in many of our partnership programs.

Eastman Scholars MathElites (ESM)

Many business leaders and their organizations are now moving away from just writing checks and toward work on systemic reforms, especially those that target mathematics and science teachers. For example, the Eastman Scholars MathElites (ESM) program, formerly called Eastman Scholars Mathletes, is one of the longest running partnerships in the nation designed to address the changing and more challenging *CCSS Mathematics*. The program promotes alignment with the revised standards by increasing the content and pedagogical knowledge of teachers, raising the number of teachers participating in standards-based professional development, providing training on standards-based resources and materials, and increasing the number of highly qualified mathematics teachers. This partnership is supported by Eastman Chemical Company, ETSU CEMSE, and school districts throughout Northeast Tennessee. Since its inception in 2007, it has prepared approximately 700 K–12 mathematics teachers by sponsoring intensive professional development delivered by ETSU CEMSE faculty during the summer months.

This training is sustained by districtwide math coaches during the academic year, and ongoing programs are given additional support through continued training on standards-based resources. To date, the ESM has seen an infusion of more than $1.3 million by its partners. This funding was especially helpful during the early years of the program when Tennessee underwent drastic changes to curriculum standards. At the start of the project in 2007, Tennessee had just received an F in the category of *Truth in Advertising* when comparing proficiency on Tennessee assessments to proficiency as determined by the National Assessment of Educational Progress (NAEP). The state also received an F in the category *Postsecondary and Workforce Readiness* (U.S. Chamber of Commerce 2007; see also Schafer, Liu, and Wang 2007; Stoneberg 2007). The ESM project was perfectly timed to make a difference, and the value of the program continues to grow as it achieves the goals established by the partnership. Other partners in our region, including Nuclear Fuels and the Niswonger Foundation, have also used the ESM model to play a significant role in their support of training for math teachers.

The ESM teachers also participate in the Upper East Tennessee Council of Teachers of Mathematics (UETCTM), which is a local affiliate of the National Council of Teachers of Mathematics (NCTM). Each ESM graduating class of teachers is given a one-year membership in the UETCTM. This membership allows math coaches and teachers to project a voice beyond the boundaries of the classroom. Starting with volume 9 in 2008, graduates of the ESM program have published essays in every volume of the UETCTM newsletter. The essays address issues called for in the *NGSS* by providing examples of ways in which teachers and math coaches create lessons and learning environments that foster students' abilities to "construct viable arguments in mathematics and critique the arguments of others" (NGAC and CCSSO 2010, pp. 6–7). The newsletter is currently on volume 17 and as many as 60 of these essays have appeared in a single volume. These newsletters are available for download at: *www.uetctm.org.*

The ESM model and its attributes have been described in greater detail by Nivens et al. (2012), and its application is not limited to math courses. The model is especially useful in science classes because teachers need training to help students engage in argumentation, use models as evidence, construct explanations using evidence and logic, and evaluate and communicate information (NRC 2012).

Eastman ScienceElites

The Eastman ScienceElites (ESE) project is an outgrowth of a professional development infrastructure for K–12 math and science teachers supported by members of ETSU CEMSE partnerships that has been around for over 25 years. CESME and its partners have trained more than 2000 inservice science teachers through training options ranging from six-week programs during the school year to two-week programs during the summer months that include sustaining support during the academic year (Rhoton and McLean 2008). Thanks to grant funding and contributions from the Niswonger Foundation, Nuclear Fuels, Wellmont Health System, Domtar Paper Company, Eastman Chemical Company, and informal science centers throughout the region, millions of dollars have been raised to support the implementation of standards-based teaching methodologies and access to standards-based resources.

The Eastman ScienceElite partnership between Eastman Chemical Company, ETSU's CEMSE, and local school districts began in 2015, and it provides upper elementary science teachers with professional development to more effectively implement the *NGSS.* This training shifts focus away from the mere delivery of science content toward using strategies that engage students in science and engineering practices and three-dimensional learning. Teachers create lesson plans according to the 5E Instructional Model (Bybee 2015) to address challenging standards and present them during the final days of the training workshop. Often, they incorporate interactive notebooks (Chesbro 2006) and integrate literacy, which are themes of the professional development. Training includes a tour of research laboratories in which teachers learn about the goals, challenges, and equipment related to different facets of STEM-related industries. To help teachers translate these experiences into learning activities appropriate for elementary students, they are asked to simulate the function of various labs through activities such as materials analysis and quality control.

The instructional team for this two-week summer workshop includes science and education faculty from ETSU, scientists and engineers from Eastman, and guest speakers from *You Be the Chemist*, a science education initiative sponsored by the Chemical Education Foundation. The collaboration by instructors mirrors that among the teacher participants who form a professional network across the region sharing electronic teaching and planning resources. Each participant receives a materials kit (supplied by Eastman) to support the implementation of learning activities from the workshop and has the opportunity to form a yearlong partnership with local scientists through the American Chemical Society.

Innovation Academy of Northeast Tennessee

Our long-term business, education, and community partnerships created a collaborative atmosphere that led to the development of a STEM-focused school. Innovation Academy of Northeast Tennessee (IA) is a groundbreaking STEM middle school that came into existence in 2012 through a partnership between the Tennessee STEM Innovation Network (TSIN), Battelle Memorial Institute, ETSU's CEMSE, local school districts, multiple business entities, community groups, and the informal science education community. The school, operated by the Sullivan County Board of Education, covers grades 6–8 and draws from assets in the community to actively engage students in learning strategies aligned to the *NGSS*.

To better achieve its vision, the school's founding principal, Sandy Watkins, and faculty developed a series of curricular units that facilitate three-dimensional learning by integrating knowledge and skills across multiple disciplines and grade spans. A key element of this curriculum is the removal of barriers that traditionally separate disciplines and the purposeful inclusion of the arts, language studies, and social sciences as partners in the STEM equation. Multiple community partnerships continue to galvanize the curriculum and provide ongoing support in the form of service performed by STEM professionals, mentoring relationships, small grants, science equipment, preservice science teacher internships, and more.

Informal Science Partnerships and *NGSS*

An eclectic array of informal science entities such as museums, nature centers, planetariums, aquariums, and zoos regularly engage learners of all ages in activities that improve their understanding of science and advance STEM literacy. A lot of these informal science-knowledge providers help K–12 to meet learning outcomes aligned with the objectives in use at most schools (Falk and Needham 2011). According to Falk and Dierking (2002), informal science experiences can play an important role in supporting the acquisition of skills, practices, and knowledge that support student success in science. Many informal science providers have a long and devoted history of working with K–12 teachers and educators to create resources, professional development, online teacher supports, and other aspects of infrastructure that multiply the effects of STEM instruction. Informal science experiences can provide novel and relevant contexts that support and enhance student engagement in the eight practices of the *NGSS* (Bell et al. 2009). The ETSU CEMSE partnerships have worked with informal education providers in our region to align their activities with the vision of the *Framework* and the *NGSS*. These informal learning experiences stimulate and engage school-age learners'

curiosity about natural phenomena through free-choice science learning.

The partnership between ETSU and the General Shale Brick Natural History Museum (the ETSU GSBNHM) in Gray, Tennessee, is an example of an alliance between an informal science center and local educators. Thousands of school-age students and educators study at the museum and fossil site each year. According to the museum's website, it "is dedicated to understanding, preserving and interpreting biodiversity of the Southern Appalachians through time, using an interdisciplinary approach. The Gray Fossil Site covers more than five acres; a Miocene site dated at 4.5 to 7 million years old, containing fossilized remains of an entire ecosystem of plants and animals. A series of sinkholes, formed from a collapsed cave, created a watering hole that drew animals from near and far. For some unlucky animals, the sinkhole became a trap that preserved them as fossils for the ages."

With support from our partners, the fossil site provides activities that build and support school-age students' interest in science, which is an attribute that some feel has been left out of the *NGSS* (Boe 2011; Lyons et al. 2012). Using the *NGSS* category Ecosystem: Interactions, Energy, and Dynamics for guidance, the GSBNHM has created a professional development system to help K–12 teachers interpret and evaluate students' understanding of ecosystem topics, beginning in the early grades. During this training, the teachers receive science and pedagogical content knowledge to help them engage students in crosscutting concepts about ecosystems across the science curriculum. To encourage lifelong learning, the museum's offerings promote STEM learning practices and a deep understanding of ecosystem ideas by engaging students in building models and describing phenomena, concepts that they can use both in and outside the school environment. The museum also offers badge-earning programs and overnight stays for Girl and Boy scouts.

Partnerships and Local STEM Conferences

A decade ago, ETSU's CEMSE recognized the importance of local STEM conferences as a means for initiating conversations about essential knowledge and capabilities that promote STEM education. These annual conferences have brought our partnerships and stakeholders together in a forum that hosts a continuing dialogue about initiatives for strengthening our work. K–12 and higher-education instructors, business and industry representatives, government entities, and other stakeholders contribute to the conversation. Conference topics have covered a wide range of topics, including navigating the *NGSS* and *CCSS*, best practices for effectively delivering STEM curriculum models, communicating workforce needs, securing and maintaining the undergraduate pipeline, government policy and its effect on the STEM teacher, listening to and supporting the concerns of teachers, and examining ways to work collaboratively to solve challenges in STEM education.

An outgrowth of our dialogue with our many partners has been a wealth of programs and initiatives that support the work of teachers, students, schools, and districts to align STEM instruction with the *NGSS*. One such example is the Governor's School in Scientific Models and Data Analysis, which allows high school students to participate in courses, laboratories, projects, field trips, seminars, lectures, and other activities centered on mathematics, statistics, and biology. A similar program, the Tennessee Junior Academy of Science (TJAS), allows high school students to engage in original science research. The success of these programs led to the development of the Middle School Summer Math and Science Scholars Program, which actively involves students in the inquiry process by allowing them to ask questions, describe objects and events, test their ideas, and communicate what they have learned through a scientific poster. These programs all provide opportunities for students to engage in science and engineering practices, such as oral discourse and argumentation from evidence, that are common across many disciplines and standards.

Final Thoughts

ETSU's CEMSE has had success building and nurturing an array of partnerships in our region to advance STEM education. As a result, we are developing a student-learning culture that is more closely aligned with the vision of the *Framework* and the *NGSS*. Programs resulting from these partnerships have led to collaboration between institutions of higher education and school districts that support long-term improvement strategies, ongoing professional development related to both content and pedagogy, access to classroom resources aligned

with core curriculum, and learning communities for students outside of formal school settings.

Partnerships play critical roles in creating and maintaining an infrastructure that supports math and science learning. Though we have received substantial financial support from our partnerships, we do not consider funding to be the main factor behind their success. Oftentimes, a financial contribution is a one-time act, and the only obligation from the contributing organization is to include the information in a report at the end of the year. Long-term STEM education partnerships can be difficult to build and even more challenging to maintain over time. Even though various members of the partnership may have common interests, it is not always easy to channel the knowledge, expertise, and resources of each partner to solve common problems. Organizers can work through these difficulties and achieve a long-term partnership by taking the time to ensure that all members experience direct benefits from the relationship.

Nurturing personal relationships in a partnership takes a lot of work and can be complicated by personnel changes over time. Fortunately, the benefits that can be gained from long-term STEM education partnerships are well worth the effort required (for example, business partners can enlist additional groups to support our efforts with whom we would otherwise never have been in contact). Communication is vital, and the goals of the partnership must be clearly and collaboratively established. All parties must agree on answers to such fundamental questions as *What do we want to accomplish? What are the metrics? How is the corporate investment going to be measured?* and *What are the expected outcomes?*

To maintain cordial relationships, it is best to avoid putting a lot of demands on the individual partners, offering them whatever latitude they want or need. Additionally, the importance of establishing personal relationships cannot be overemphasized. There is no substitute for meeting face-to-face with potential partners to see if they share your vision of STEM education and what they can offer to support it. Once these relationships have been established, it is important to maintain them for the length of the partnership.

References

Bell, P., B. Lewenstein, A. Shouse, and M. A. Feder. 2009. *Learning sciences in informal environments: People, places* _and pursuits._ Washington, DC: National Academies Press.

Boe, M.V. 2011. Science choices in Norwegian upper secondary school: What matters? _Science Education_ 96 (1): 1–20

Bybee, R. W. 2015. _The BSCS 5E Instructional model: Creating teachable moments._ Arlington, VA: NSTA Press.

Chesbro, R. 2006. Using interactive science notebooks for inquiry-based science. _Science Scope_ 29 (7): 30–34.

Falk, J. H., and L.D. Dierking. 2002. _Lessons without limit: How free-choice learning is transforming education._ Lanham, MD: AltaMira Press.

Falk, J. H., and M. Needham. 2011. Measuring the impact of a science center on its community. _Journal of Research in Science Teaching_ 48 (1): 1–12.

Lightbown, P. M., and N. Spada. 2006. _How languages are learned._ 3rd ed. Oxford, UK: Oxford University Press.

Lyons, T., Quinn, F., Rizk, N., Anderson, N., Hubber, P., and J. Kenny. 2012. _Starting out in STEM: A study of young men and women in first year science, technology, engineering and mathematics courses._ Armidale, New South Wales, Australia: SiMEER National Research Center.

National Research Council (NRC). 2012. _A framework for K–12 science education: Practices, crosscutting concepts, and core ideas._ Washington, DC: National Academies Press.

National Governors Association Center for Best Practices and Council of Chief State School Officers (NGAC and CCSSO). 2010. _Common core state standards._ Washington, DC: NGAC and CCSSO.

NGSS Lead States. 2013. _Next Generation Science Standards: For states, by states._ Washington, DC: National Academies Press. _www.nextgenscience.org/ next-generation-science-standards._

Nivens, R. A., J. Rhoton, G. Poole, H. Imboden, T. Peters, and S. Harvey. 2012. The Eastman Scholar Mathletes: A collaborative partnership. _Tennessee Association of Middle Schools Journal_ Spring: 93–102.

Pearson, P. D., E. Moje, and C. L. Greenleaf. 2010. Literacy and science: Each in the service of the other. _Science_ 328 (5977): 459–463.

Rhoton, J., and J. E. McLean. 2008. Developing teacher leaders in science: Catalysts for improved science teaching and student learning. _Science Educator_ 17 (2): 45–56.

Schafer, W. D., M. Liu, and H. J. Wang. 2007. Content and grade trends in state assessments and NAEP. _Practical Assessment Research & Evaluation_ 12 (9): 1–25 _http:// pareonline.net/getvn.asp?v=12&n=5._

Stoneberg, B. D. 2007. Using NAEP to confirm state test results in the No Child Left Behind. _Practical Assessment Research & Evaluation._ 12 (5): 1–10. _http:// pareonline.net/getvn.asp?v=12&n=5._

U.S. Chamber of Commerce. 2007. _Report card 2007: Overview and map._ Washington, DC: U.S. Chamber of Commerce. _www.uschamber.com/report/ report-card-2007-overview-and-map._

Wallace, C. S. 2004. An illumination of the roles of hands-on activities, discussion, text reading, and writing in constructing biology knowledge in seventh grade. _School Science and Mathematics_ 104 (2): 1–9.

Chih-Che Tai _is assistant director of the Center of Excellence in Mathematics and Science Education and an associate professor of science education at East Tennessee State University. He is actively involved in grant activities to improve math and science education through partnerships with local educational agents, K–12 teachers, university faculty, and business partners. His research interests include integrating science technology, engineering technology, and literacy into K–12 education, as well as college and career readiness, teacher professional development, and teacher effectiveness. Tai holds a BS and MS in chemistry from National Taiwan University and a PhD in science education from Teachers College, Columbia University._

Ryan Nivens _is an associate professor of mathematics education at East Tennessee State University and is extensively involved in undergraduate and graduate STEM education. He is an instructor in the Eastman Chemical Company's Scholar MathElites program and was a consultant for the State of Missouri Mathematics Academy for seven years. Since 2008, he has authored nine STEM grants, which brought in $228,000, and he has coauthored five STEM grants, which have secured more than $500,000. He has served as president of two mathematics education organizations and has published several articles on innovative ways to teach mathematics. His current involvement is influenced by his early career experiences working as a secondary mathematics teacher._

Laura Robertson *is an assistant professor of science education at East Tennessee State University. She teaches science content and methods courses for preservice teachers and leads professional development workshops. Her research interests include teacher autonomy, professional learning communities, and the integration of science and English language arts. Before serving in her current position, Robertson was a middle school science teacher for 11 years.*

Karin Keith *is associate professor and department chair for the Department of Curriculum and Instruction at East Tennessee State University. Before this, she was a literacy coach in Johnson City Schools, where she initiated a district coaching model. Keith has authored several journal articles and chapters as well as served as a reviewer for the* Journal of Teacher Education *and the* Literacy Research Association.

Anant Godbole *is a mathematical scientist who has held positions at East Tennessee State University, Michigan Technological University, University of California (Santa Barbara and Berkeley), and Johns Hopkins University. He has been department chair and associate dean and is currently director of the ETSU Center of Excellence in STEM Education and of the Northeast Tennessee STEM Hub. Godbole has written over more than professional papers, and he has been awarded around $8 million in grant funding, mainly from the National Science Foundation. Additionally, he has run a nationally recognized Research Experiences for Undergraduates program for 25 years.*

Jack Rhoton *is professor emeritus and former executive director of the Center of Excellence in Mathematics and Science Education at East Tennessee State University.*

Index

Page numbers printed in **boldface type** indicate tables or figures.

Index

Index

and case-based instruction, 121
claims-evidence-reasoning (CER) framework, 12, 15, 16, 71, 72
distractor test-item alternatives, 72
instructing students in, 7, 8
modeling to support development of, 72–73
and phenomena-based investigations, 31–33, **31**, **32**
presentation of, 26
science talk moves, 70–71, **71**

F

federal policies, *NGSS* and, 88
5E Instructional Model, 6–7, 25, 109, 112, 156
Five Tools and Processes for Translating the *NGSS*, 55–56
foreground content, 6–8
formative assessment
    adding argumentation to, 42–44, **43**
    creating tasks for, 44–46
    crosscutting concepts and, 40–42, **40, 41**
    probes as, 71–72
    rubrics and, 75
A Framework for K–12 Science Education, 3, 5, 29–30, 39–40, 60

G

General Shale Brick Natural History Museum, 157
Generate an Argument model, 100–102
global climate change inquiry
    about, 133–135
    evidence gathering, 135–136
    evidence understanding, 136–141, **137, 138, 139**
    impact of *NGSS* phenomena approach, 142
    importance of phenomena-based investigations, 134–135
    information for students, 141–142
Gottesman Center for Science Teaching and Learning at the American Museum of Natural History, 55
Governor's School in Scientific Models and Data Analysis, 158
graduation policies, 89
graphic organizers, 16
Guidelines for the Evaluation of Instructional Materials for Science (BSCS), 55

A Guide to Implementing the Next Generation Science Standards, 60

H

habits of mind, 5, 9
higher education, roles and responsibilities, 147–149
higher-order cognitive skills, 17

I

informal science experiences, 157
Innovation Academy of Northeast Tennessee, 154, 156–157
inquiry, 3–4, 24–25, 31
instructional materials
    importance, 52
    teacher selection of, 54–55
    tools that support, 53–54
interdisciplinary instruction, 5

J

joy of learning, 9

K

K–12 Alliance, 54–55, 62, 63
K–12 STEM education, 146–149

L

Learning Cycles, 121
learning environments, 120
lesson study, 63, 65

M

Mathematical Knowledge for Teachers (MKT), 109
mathematics, 5
meaningful learning, 9. *See also* real-life connections
Mid-Continent Research for Education and Learning (McREL), 39
Mini Think Tanks, 35
modeling, scientific, 72–73, 82, 127
museums, 157

Index

Index